Gianluca Crippa, Anna Mazzucato (Eds.)

Transport, Fluids, and Mixing

Gianluca Crippa, Anna Mazzucato (Ed.)

Transport, Fluids, and Mixing

Managing Editor: Agnieszka Bednarczyk-Drąg
Series Editor: Gianluca Crippa
Language Editor: Adam Tod Leverton

Open Access
Partial Differential Equations and Measure Theory

DE GRUYTER
OPEN

ISBN 978-3-11-057123-3
e-ISBN (PDF) 978-3-11-057124-0
e-ISBN (EPUB) 978-3-11-057146-2

Library of Congress Cataloging-in-Publication Data
A CIP catalog record for this book has been applied for at the Library of Congress.

Published by De Gruyter Open Ltd, Warsaw/Berlin
Part of Walter de Gruyter GmbH, Berlin/Boston
The book is published with open access at www.degruyter.com.
Cover illustration: ©samsonovs

Managing Editor: Agnieszka Bednarczyk-Drąg
Series Editor: Gianluca Crippa
Language Editor: Adam Tod Leverton

www.degruyteropen.com

Contents

Gianluca Crippa and Anna Mazzucato

Introduction

This volume contains the lecture notes originating from the Summer School "Transport, Fluids, and Mixing" held in Levico Terme, Italy, July 19–24, 2015, organized by the Editors with the support and cooperation of CIRM - Centro Internazionale per la Ricerca Matematica (International Center for Mathematics Research) and of the FBK - Fondazione Bruno Kessler (Bruno Kessler Foundation).

The main scope of the summer school was to introduce junior researchers, graduate students and postdocs, to the state of the art in the analysis of transport and mixing phenomena in fluids and associated problems. The present volume contains an abridged version of the school short courses, together with introductory material and recent developments. It is aimed at a broad audience of mathematicians working in partial differential equations (PDEs) and applied mathematics with an interests in fluid mechanics and related areas. The volume supplements the current literature with a unique blend of contributions that touch upon both theoretical as well as modeling questions and present a variety of techniques, from the analysis of PDEs, to harmonic analysis, to computational methods. Each chapter in the volume has been written by one of the main lecturers of the Levico summer school with the assistance of a few of the junior participants, postdocs and graduate students.

Recent progress in addressing some of the most outstanding open questions in the mathematical analysis of fluid flows, in particular finite-time blow up for strong solutions and non uniqueness of weak solutions to the Euler and Navier-Stokes equations, and Onsager's Conjecture on energy dissipative solutions to the Euler equations, has been brought about by a sometimes novel, often renewed, interest in combining traditional approaches to solving PDEs, such as energy estimates, with techniques developed in other contexts, as in geometric and harmonic analysis and in probability. An especially poignant example of this cross-fertilization comes from the successes of *convex integration*. Initially developed in geometric analysis as a way to prove the homotopy or *h-principle* for differential relations (the reader is referred to the survey article [24] and references therein), it has proven instrumental in establishing some of the most important results concerning the Euler equations, namely, non-uniqueness for weak solutions [23] and Onsager's Conjecture [12, 13, 18, 30], which has fundamental implications for a mathematical theory of turbulence.

Transport and diffusion represent complementary, and sometimes competing, phenomena at the core of any modeling of incompressible fluid flows. From a mathe-

Gianluca Crippa: Departement Mathematik und Informatik, Universität Basel, Spiegelgasse 1, CH-4051, Basel, Switzerland, E-mail: gianluca.crippa@unibas.ch
Anna Mazzucato: Department of Mathematics, Penn State University, University Park, PA, 16802, U.S.A., E-mail: alm24@psu.edu

matical point of view, *irregular transport* by non-smooth velocity fields has been recognized as an important mechanism that can sustain *small-scale creation*, enhancing *dissipation* and *mixing* of quantities advected by the flow. While the theory of solutions to transport equations with Lipschitz velocity can be reduced to the corresponding theory for ordinary differential equations (ODEs) and it is classical, there are still many open questions regarding the behavior of solutions when the velocity has only Sobolev regularity in space with an integral dependence on time. In this context, the appropriate notion of solution is that of a *renormalized solution*, introduced in the seminal work of DiPerna-Lions [26] and extended to velocities with bounded variation by Ambrosio [4]. A related concept is that of a *regular Lagrangian flow*, which leads to a quantitative theory for solutions to ODEs with non-Lipschitz velocity [22]. The work of Crippa and De Lellis in [22] was motivated by a conjecture of Bressan on the cost of rearrangements [11], and has lead to significant advances in establishing rigorous bounds on mixing rates of passive scalars advected by incompressible flows.

Quantifying mixing has proven a rich source of problems in a variety of fields. There is a well-established literature on mixing in chaotic dynamics (see e.g. [5, 38, 42]). In turbulence, advection enhances mixing, which in turn can enhance diffusion and suppress concentration (see [19] for steady "relaxation enhancing" flows and [35] for an application to chemotaxis, for instance), with implications for statistical properties (see e.g. [8, 28] and references therein) of turbulent flows. Enhanced dissipation can occur also in Euler flows as an effect of inviscid damping (see [7] and references therein). A quantitative measure of mixing is the decay in time of spatial correlations. In the absence of diffusion, covariance of the passive scalar field is no longer a viable measure of mixing. Negative Sobolev norms, which penalize large scales, provide a multi-scale measure of mixing, which can be readily monitored in applications. Decay in time of these norms is related to strong mixing in the ergodic sense by the underlying flow, as observed by Mathew, Mezič, and Petzold [40], and identifies a characteristic scale, which serves as a "functional mixing length". A related "geometric mixing length" arises naturally from Bressan's conjecture. The sharpness of the decay rates in time for the mixing lengths under physically-based constrained on the velocity, such as energy, enstrophy, or palenstrophy budgets, has been investigated both computationally, as well as analytically, in particular in the work of Doering, Thiffault, and collaborators first [37, 39], and then by Iyer, Kiselev, Xu [31] and Seis [44], using a variety of techniques, from energy estimates, to geometric measure theory, to optimal transport, to multilinear harmonic analysis [36] (for related results in the context of singular perturbation analysis, see Otto and collaborators [41, 43]). Examples of optimal mixers have been recently constructed by Yao and Zlatos [46], and by Alberti, Crippa, and Mazzucato [2, 3], utilizing both analytic as well as geometric approaches.

Chapter 1, *Lectures on Stirring, Mixing and Transport* (C. Doering, in collaboration with C. Nobili), opens the volume with an introduction to the concepts of mixing and transport of a passive tracer due to stirring by an incompressible flow. The problem whether advection can enhance dissipation via particle dispersion is then examined, together with concepts of mixing efficiency. A discussion of mixing and transport in the presence of sources and sinks closes the chapter.

Renormalized solutions to transport equations can be viewed in the more general context of *generalized flows* and *measure-valued solutions* or *Young measures*, which can capture the defects produced by oscillations and concentration along weakly convergent approximating sequences in non-linear equations. Starting with the seminal work by DiPerna and Majda [27], measure-valued solutions have been employed to study energy concentration for solutions to the Euler equations. Recently, an application of convex integration has led to the construction of Young measures that yield global-in-time weak solutions of the Cauchy problem for a large set of initial data for the 3D Euler equations [45]. These solutions are highly non unique, though they have energy that is bounded in time.

Chapter 2, *New concepts of solutions in fluid dynamics* (Y. Brenier, in collaboration with L. Keller), present new possible ways to define generalized flows and generalized solutions to the Euler equations. Motivated by the hydrostatic limit in the three-dimensional Euler equations, generalized flows are first described, which can be viewed as solutions with a microstructure that allows to restore uniqueness of trajectories and equivalence between Eulerian and Lagrangian formulations in a suitable way. Then, a new approach to generalized solutions is presented that borrows ideas from the theory of probability measures on path space. Lastly, interesting examples are given and a connection is made with the optimal transportation problem and Monge-Ampère equation.

Advection-diffusion equations arise naturally in modeling incompressible fluid flow. For instance, in two space dimensions, the Navier-Stokes equations can be recast as an active transport-diffusion equation for the vorticity. Another, more singular and challenging, equation of this type is the *surface quasi-geostrophic equation* (SQG), which has been derived in the analysis of rapidly rotating geophysical flows. There are still several open questions regarding the well posedness of SQG and properties of its solutions, though much progress has been achieved recently, again by exploiting an array of different techniques. Indeed, global existence for the viscous critical SQG was obtained independently by Caffarelli and Vasseur [15], using a Nash-Moser iteration, and by Kiselev, Nazarov, and Volberg [33], using the evolution of an appropriate modulus of continuity (see [21] for a third proof that uses a non-linear maximum principle). On the other hand, even in the viscous case, non-uniqueness of suitable weak solutions holds by means of convex integration as shown by Buckmaster, Shkoller, and Vicol [14].

Chapter 3, *Existence, uniqueness, regularity and long time behavior of hydrodynamic evolution equations* (P. Constantin, in collaboration with L. Keller and C. Nobili), presents some recent results concerning the well posedness of SQG and a related problem, arising from coupling the Navier-Stokes equations with other fields equations, such as in magneto-hydrodynamics or in modeling Oldroyd-B-type complex fluids. Existence and uniqueness for these systems is established by combining

PDE techniques, in particular an Eulerian-Lagrangian approach to the equations of motion (introduced by Constantin in [16, 17]) and Nash iteration, together with harmonic analysis tools, more specifically commutator estimates for singular integrals and a non-linear lower bound on the fractional Laplacian [21]. The dependence in SQG of the advecting velocity on the advected quantity via a singular integral makes the use of harmonic analysis particularly appealing.

Determining finite-time blow-up for strong solutions and regularity of weak solutions for both the Euler and Navier-Stokes equations remains one of the most outstanding open problems in the analysis of PDEs and, arguably, mathematics as a whole. One of the main mechanisms for possible blow-up of strong solutions to the Euler equations in three space dimensions is growth of vorticity (by the celebrated Beale-Kato-Majda criterion [6]), where growth can be strongly localized spatially. Even in two space dimensions, where a uniform bound on vorticity holds, the gradient of vorticity can grow unboundedly, signaling the creation of small scales in the flow. A double exponential-in-time bound on the Lipschitz norm of vorticity can be established *a priori* and a natural question is whether this bound is sharp. Examples where the growth is superlinear can be found in the literature [25]. For linear equations, filamentation and mixing due to irregular transport can also produce growth of positive Sobolev norms, which can lead to loss of regularity for the solution [1]. As a matter of fact, Jabin [32] exhibits an example of a discontinuous flow in Sobolev spaces with $W^{1,p}$ velocity field. At the same time, some very low regularity is retained, a fact that has been used for a new theory of existence of solutions to the compressible Navier-Stokes equations [10].

Chapter 4, *Small scale creation in inviscid fluids* (A. Kiselev, in collaboration with M. Chernobay, O. Lazar, and C. Li) closes the volume, presenting recent results on the growth of vorticity and its gradient in respectively two and three space dimensions and on the regularity of solutions to the Euler equations. The chapter opens with a review of known results on well posedness of the two-dimensional Euler equations and a review of available *a priori* bounds on solutions. After surveying examples of flows that lead to growth of quantities passively advected, in particular the so-called Bahouri-Chemin flow [9], the recent result by Kiselev and Svérak [34], establishing the sharpness of the double exponential growth of the vorticity gradient, is discussed. Lastly, reduced models for the three-dimensional Euler equations exhibiting finite-time blow up are presented, from classical models such as the Constantin-Lax-Majda model [20] to more recent models, such as the Hou-Luo model [29].

To conclude, this volume addresses both qualitative and quantitative aspects of mixing and transport in fluid flows from the point of view of mathematical analysis, and brings together leading experts from diverse areas that can benefit from cross fertilization. Transport and mixing are important phenomena in many applied fields, from environmental science, to geophysics and climate modeling. As such, it is expected that the volume will be a useful reference for applied mathematicians working on interdisciplinary problems.

Acknowledgment: The Editors gratefully acknowledge the fundamental contributions of the main summer school lecturers and authors, Yann Brenier, Peter Constantin, Charles Doering, and Alexander Kiselev, and their co-authors Michail Chernobay, Laura Keller, Omar Lazar, Chao Li, and Camilla Nobili, to the success of this lecture notes volume. They also acknowledge the support of CIRM-FBK, the University of Basel, the US National Science Foundation, Penn State University, the A.P.T. Valsugana and the Comune di Levico Terme. Without their support the summer school and this volume would have not been possible. A. Mazzucato was partially supported by the US National Science Foundation grants DMS 1009713, 1009714, and 1312727.

This publication in digital format has been produced with the support of the Swiss National Science Foundation. This publication in printed format has been supported pre-press by the Swiss National Science Foundation. Die digitale Version wurde publiziert mit Unterstützung des Schweizerischen Nationalfonds zur Förderung der wissenschaftlichen Forschung. Die Druckvorstufe der Druckversion dieser Publikation wurde vom Schweizerischen Nationalfonds zur Förderung der wissenschaftlichen Forschung unterstützt.

Bibliography

[1] G. Alberti, G. Crippa, and A. L. Mazzucato. Loss of regularity for the continuity equation with non Lipschitz velocity field. In preparation.
[2] G. Alberti, G. Crippa, and A. L. Mazzucato. Exponential self-similar mixing and loss of regularity for continuity equations. *C. R. Math. Acad. Sci. Paris*, 352(11):901–906, 2014.
[3] G. Alberti, G. Crippa, and A. L. Mazzucato. Exponential self-similar mixing by incompressible flows. *ArXiv e-prints*, May 2016.
[4] L. Ambrosio. Transport equation and Cauchy problem for *BV* vector fields. *Invent. Math.*, 158(2):227–260, 2004.
[5] H. Aref. Stirring by chaotic advection. *J. Fluid Mech.*, 143:1–21, 1984.
[6] J. T. Beale, T. Kato, and A. Majda. Remarks on the breakdown of smooth solutions for the 3-D Euler equations. *Comm. Math. Phys.*, 94(1):61–66, 1984.
[7] J. Bedrossian, N. Masmoudi, and V. Vicol. Enhanced dissipation and inviscid damping in the inviscid limit of the Navier-Stokes equations near the two dimensional Couette flow. *Arch. Ration. Mech. Anal.*, 219(3):1087–1159, 2016.
[8] G. Boffetta, A. Celani, M. Cencini, G. Lacorata, and A. Vulpiani. Nonasymptotic properties of transport and mixing. *Chaos*, 10(1):50–60, 2000.
[9] F. Bouchut and G. Crippa. Lagrangian flows for vector fields with gradient given by a singular integral. *J. Hyperbolic Differ. Equ.*, 10(2):235–282, 2013.
[10] D. Bresch and P.-E. Jabin. Global existence of weak solutions for compressible Navier-Stokes equations: Thermodynamically unstable pressure and anisotropic viscous stress tensor. Preprint 2015.
[11] A. Bressan. A lemma and a conjecture on the cost of rearrangements. *Rend. Sem. Mat. Univ. Padova*, 110:97–102, 2003.
[12] T. Buckmaster, C. De Lellis, and L. Székelyhidi, Jr. Dissipative Euler flows with Onsager-critical spatial regularity. *Comm. Pure Appl. Math.*, 69(9):1613–1670, 2016.

[13] T. Buckmaster, C. De Lellis, L. Székelyhidi, Jr., and V. Vicol. Onsager's conjecture for admissible weak solutions. *ArXiv e-prints*, Jan. 2017.

[14] T. Buckmaster, S. Shkoller, and V. Vicol. Nonuniqueness of weak solutions to the SQG equation. *ArXiv e-prints*, Oct. 2016.

[15] L. A. Caffarelli and A. Vasseur. Drift diffusion equations with fractional diffusion and the quasi-geostrophic equation. *Ann. of Math. (2)*, 171(3):1903–1930, 2010.

[16] P. Constantin. An Eulerian-Lagrangian approach for incompressible fluids: local theory. *J. Amer. Math. Soc.*, 14(2):263–278 (electronic), 2001.

[17] P. Constantin. An Eulerian-Lagrangian approach to the Navier-Stokes equations. *Comm. Math. Phys.*, 216(3):663–686, 2001.

[18] P. Constantin, W. E, and E. S. Titi. Onsager's conjecture on the energy conservation for solutions of Euler's equation. *Comm. Math. Phys.*, 165(1):207–209, 1994.

[19] P. Constantin, A. Kiselev, L. Ryzhik, and A. Zlatoš. Diffusion and mixing in fluid flow. *Ann. of Math. (2)*, 168(2):643–674, 2008.

[20] P. Constantin, P. D. Lax, and A. Majda. A simple one-dimensional model for the three-dimensional vorticity equation. *Comm. Pure Appl. Math.*, 38(6):715–724, 1985.

[21] P. Constantin and V. Vicol. Nonlinear maximum principles for dissipative linear nonlocal operators and applications. *Geom. Funct. Anal.*, 22(5):1289–1321, 2012.

[22] G. Crippa and C. De Lellis. Estimates and regularity results for the DiPerna-Lions flow. *J. Reine Angew. Math.*, 616:15–46, 2008.

[23] C. De Lellis and L. Székelyhidi, Jr. The Euler equations as a differential inclusion. *Ann. of Math. (2)*, 170(3):1417–1436, 2009.

[24] C. De Lellis and L. Székelyhidi, Jr. The h-principle and the equations of fluid dynamics. *Bull. Amer. Math. Soc. (N.S.)*, 49(3):347–375, 2012.

[25] S. A. Denisov. Infinite superlinear growth of the gradient for the two-dimensional Euler equation. *Discrete Contin. Dyn. Syst.*, 23(3):755–764, 2009.

[26] R. J. DiPerna and P.-L. Lions. Ordinary differential equations, transport theory and Sobolev spaces. *Invent. Math.*, 98(3):511–547, 1989.

[27] R. J. DiPerna and A. J. Majda. Oscillations and concentrations in weak solutions of the incompressible fluid equations. *Comm. Math. Phys.*, 108(4):667–689, 1987.

[28] T. Gotoh and T. Watanabe. Scalar flux in a uniform mean scalar gradient in homogeneous isotropic steady turbulence. *Phys. D*, 241(3):141–148, 2012.

[29] T. Y. Hou, Z. Lei, G. Luo, S. Wang, and C. Zou. On finite time singularity and global regularity of an axisymmetric model for the 3D Euler equations. *Arch. Ration. Mech. Anal.*, 212(2):683–706, 2014.

[30] P. Isett. A Proof of Onsager's Conjecture. *ArXiv e-prints*, Aug. 2016.

[31] G. Iyer, A. Kiselev, and X. Xu. Lower bounds on the mix norm of passive scalars advected by incompressible enstrophy-constrained flows. *Nonlinearity*, 27(5):973–985, 2014.

[32] P.-E. Jabin. Critical non-Sobolev regularity for continuity equations with rough velocity fields. *J. Differential Equations*, 260(5):4739–4757, 2016.

[33] A. Kiselev, F. Nazarov, and A. Volberg. Global well-posedness for the critical 2D dissipative quasi-geostrophic equation. *Invent. Math.*, 167(3):445–453, 2007.

[34] A. Kiselev and V. Šverák. Small scale creation for solutions of the incompressible two-dimensional Euler equation. *Ann. of Math. (2)*, 180(3):1205–1220, 2014.

[35] A. Kiselev and X. Xu. Suppression of chemotactic explosion by mixing. *Arch. Ration. Mech. Anal.*, 222(2):1077–1112, 2016.

[36] F. Léger. A new approach to bounds on mixing. Preprint arXiv:1604.00907.

[37] Z. Lin, J.-L. Thiffeault, and C. R. Doering. Optimal stirring strategies for passive scalar mixing. *J. Fluid Mech.*, 675:465–476, 2011.

[38] C. Liverani. On contact Anosov flows. *Ann. of Math. (2)*, 159(3):1275–1312, 2004.
[39] E. Lunasin, Z. Lin, A. Novikov, A. L. Mazzucato, and C. Doering. Optimal mixing and optimal stirring for fixed energy, fixed power or fixed palenstrophy flows. *Journal of Mathematical Physics*, 53:115611, 15 pp., 2012.
[40] G. Mathew, I. Mezić, and L. Petzold. A multiscale measure for mixing. *Phys. D*, 211(1-2):23–46, 2005.
[41] G. Menon and F. Otto. Diffusive slowdown in miscible viscous fingering. *Commun. Math. Sci.*, 4(1):267–273, 2006.
[42] J. M. Ottino. *The kinematics of mixing: stretching, chaos, and transport*. Cambridge Texts in Applied Mathematics. Cambridge University Press, Cambridge, 1989.
[43] F. Otto. Viscous fingering: an optimal bound on the growth rate of the mixing zone. *SIAM J. Appl. Math.*, 57(4):982–990, 1997.
[44] C. Seis. Maximal mixing by incompressible fluid flows. *Nonlinearity*, 26(12):3279–3289, 2013.
[45] L. Székelyhidi and E. Wiedemann. Young measures generated by ideal incompressible fluid flows. *Arch. Ration. Mech. Anal.*, 206(1):333–366, 2012.
[46] Y. Yao and A. Zlatos. Mixing and Un-mixing by Incompressible Flows. *ArXiv e-prints*, July 2014.

Charles R. Doering* and Camilla Nobili

Lectures on Stirring, Mixing and Transport

Abstract: These lectures introduce some basic concepts and mathematical approaches to the quantitative study of passive tracer mixing, dispersion and transport resulting from stirring by incompressible flows. Qualitative and quantitative similarities and differences between mixing and transport are explored. We focus on notions of *effective diffusion* and its dependence on the strength of stirring relative to molecular diffusion.

Keywords: Mixing, transport, dispersion, advection-diffusion equation, diffusion processes, stochastic differential equations, Itô calculus

1.1 Stirring, mixing and transport

The words stirring, mixing and transport are often interchanged but they describe very different processes. Let us distinguish the notions:

Stirring: the flow of a fluid.
Mixing: the intermingling of distinct materials or fluid properties that were originally separated in space.
Transport: the displacement of material or fluid properties from one place to another.

There are various levels at which mixing can be observed. At the smallest scales mixing happens via molecular diffusion, and this may be a very slow process. Molecular diffusion is described by the heat equation

$$\dot{T} = \kappa \Delta T \,, \tag{1.1}$$

where $T(x, t)$ is the concentration of a material or density or intensity of a property (such as temperature) in the fluid, and κ is the *molecular diffusion coefficient* which is a property of the solvent and the material or property that is dissolved in it. The heat equation (1.1) must also be augmented with suitable initial and boundary conditions which depend on the specific application. In these lectures we will consider T to be the density of a *passive tracer*, a concentration of particles (or markers) that do not feed back and affect the motion of the fluid.

For simplicity let $x \in \mathbb{R}$ (what follows can be easily generalized to $\boldsymbol{x} \in \mathbb{R}^n$) and suppose that at the initial time $t = 0$ the concentration T satisfies

$$T(x, 0) = \delta(x) \,,$$

***Corresponding Author: Charles R. Doering:** Center for the Study of Complex Systems, University of Michigan, U.S.A., E-mail: doering@umich.edu
Camilla Nobili: Departement Mathematik und Informatik, Universität Basel, Switzerland, E-mail: camilla.nobili@unibas.ch

where $\delta(x)$ is the delta distribution centered at $x = 0$. Then the solution is

$$T(x,t) = \frac{1}{\sqrt{4\pi\kappa t}} \exp\left(-\frac{x^2}{4\kappa t}\right).$$

We first observe that the integral over the whole space is preserved at each time

$$\int T(x,t)\,dx = 1\,, \tag{1.2}$$

and in particular $\frac{d}{dt}\int T\,dx = 0$. Moreover, since there is no drift, the center of mass-location does not move with time,

$$\int xT(x,t)\,dx = 0. \tag{1.3}$$

To see this, multiply (1.1) by x and integrate by parts:

$$\frac{d}{dt}\int xT\,dx = \int x\partial_t T\,dx = \kappa\int x\partial_x^2 T\,dx = -\kappa\int \partial_x T\,dx = 0.$$

On the other hand the width of the distribution, i.e., the variance, grows linearly in time:

$$\int x^2 T(x,t)\,dx = 2\kappa t. \tag{1.4}$$

To see this, multiply (1.1) by x^2 and integrate by parts twice:

$$\frac{d}{dt}\int x^2 T\,dx = \int x^2\partial_t T\,dx = \kappa\int x^2\partial_x^2 T\,dx = 2\kappa\int T\,dx = 2\kappa.$$

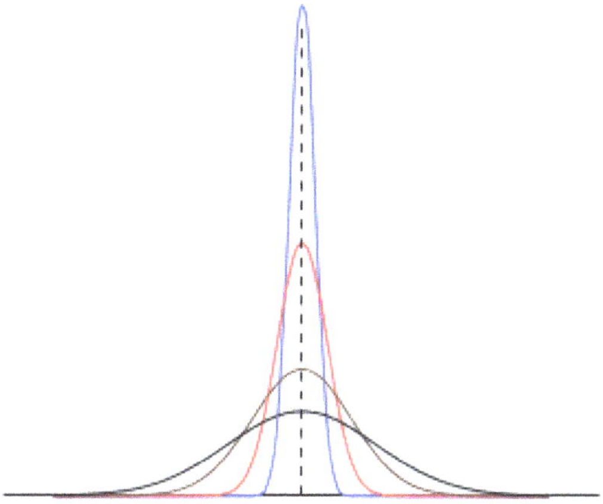

Fig. 1.1. Gaussian distributions.

Molecular diffusion coefficients are typically small numbers in familiar physical units. For thermal diffusion in water, for example, $\kappa \approx 10^{-5}\frac{m^2}{s}$ while for, say, ink in water $\kappa \approx 10^{-7}\frac{m^2}{s}$.

Now consider a stochastic process X_t describing the position of the particle at time t. Suppose that the process starts at 0 at $t = 0$, i.e., $X_0 = 0$, and subsequently evolves according to

$$dX_t = \sqrt{2\kappa}\, dW_t, \tag{1.5}$$

where W_t is the Wiener process (see Appendix). This stochastic dynamics corresponds to the heat equation in the sense that the time-dependent probability density for X_t satisfies (1.1). Using the stochastic calculus we can directly compute the expectations $\mathbb{E}(X_t)$ and $\mathbb{E}(X_t^2)$ and see that they coincide with (1.3) and (1.4).

First, since the Wiener process satisfies

$$\mathbb{E}(dW_t) = 0, \tag{1.6}$$

by (1.5) we have

$$\mathbb{E}(dX_t) = d\mathbb{E}(X_t) = \sqrt{2\kappa}\,\mathbb{E}(dW_t) = 0,$$

which implies $\mathbb{E}(X_t) = \mathbb{E}(X_0) = 0$.

Then using the Itô formula (see Appendix) to write the stochastic increment of X_t^2,

$$d(X_t^2) = 2X_t dX_t + \frac{1}{2}\, 2\,(dX_t)^2\,, \tag{1.7}$$

we have

$$\mathbb{E}(d(X_t^2)) = 2\,\mathbb{E}(X_t dX_t) + \frac{1}{2}\, 2\,\mathbb{E}((dX_t)^2). \tag{1.8}$$

The first term of the right hand side vanishes thanks to (1.5) and the property of Itô equations that, for any function $\phi(\cdot)$,

$$\mathbb{E}(\phi(X_t) dW_t) = \mathbb{E}(\phi(X_t))\,\mathbb{E}(dW_t) = 0. \tag{1.9}$$

The second term is $\mathbb{E}((dX_t)^2) = 2\kappa\mathbb{E}(dW_t^2) = 2\kappa dt$ where we call on the property of the Wiener process that

$$\mathbb{E}(dW_t^2) = dt. \tag{1.10}$$

Therefore $\mathbb{E}(d(X_t^2)) = d\mathbb{E}(X_t^2) = 2\kappa dt$ so that $\mathbb{E}(X_t^2) = 2\kappa t$.

This simple example illustrates how the stochastic differential equation formalism can, at least in some cases, produce quick and clean computations of moments.

1.2 Advection as diffusion

Consider the two-level Markov process $I_t = \pm 1$, which consists of randomly switching between the $(+1)$ and (-1) states. The waiting times between jumps are independent and exponentially distributed. If $\mathbb{E}(I_0) = 0$ then

$$\mathbb{E}(I_t I_s) = \exp(-2\gamma|t-s|) = \exp\left(-\frac{|t-s|}{\tau}\right), \tag{1.11}$$

where $\tau = \frac{1}{2\gamma}$ is the process' correlation time.

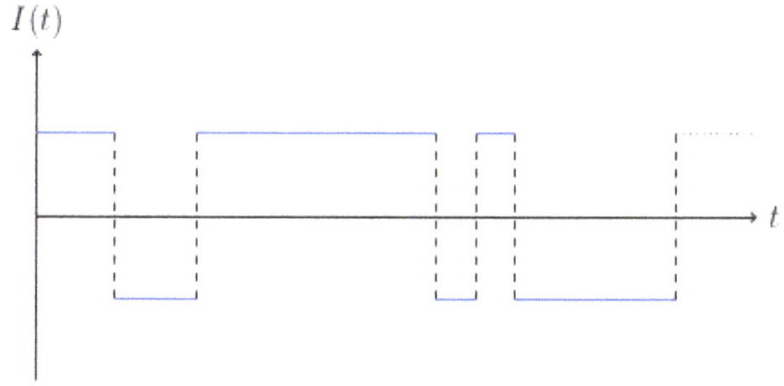

Fig. 1.2. The two-level (dichotomous) Markov process that jumps between the (±1) states.

We denote the probability of being in the (±1) state at time t by $p^{\pm}(t) \in [0, 1]$. The rate of change of the probabilities is summarized by the ODEs

$$\frac{d}{dt}\begin{bmatrix} p^+(t) \\ p^-(t) \end{bmatrix} = \begin{bmatrix} -\gamma & \gamma \\ \gamma & -\gamma \end{bmatrix} \begin{bmatrix} p^+ \\ p^- \end{bmatrix}.$$

Now consider a particle whose velocity switches randomly between $\pm U$ (where U is a real – without loss of generality positive – number) in the two-level Markov process so that its position as a function of time X_t satisfies the differential equation

$$dX_t = U I_t \, dt. \tag{1.12}$$

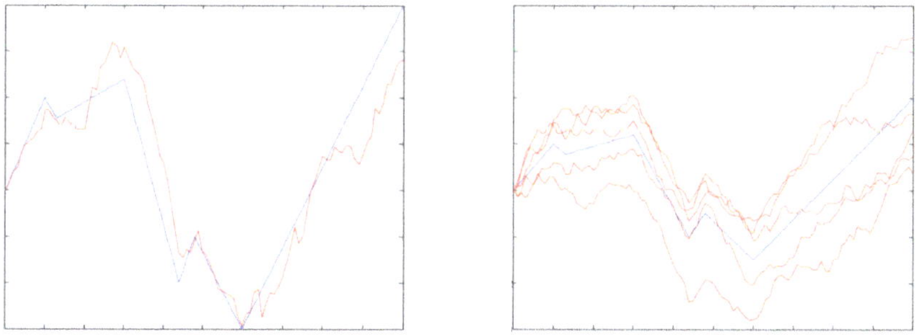

Fig. 1.3. Particle trajectories: in blue the piecewise constant velocity path and in red an illustrative Brownian motion (left) and many Brownian motions (right).

The probability densities $\rho^{\pm}(x,t)$ — the densities for the particle's position X_t conditioned on its velocity being $\pm U$ at time t — evolve according to

$$\frac{\partial}{\partial t}\begin{bmatrix}\rho^+(x,t)\\ \rho^-(x,t)\end{bmatrix} = \begin{bmatrix}-U\partial_x - \gamma & \gamma\\ \gamma & U\partial_x - \gamma\end{bmatrix}\begin{bmatrix}\rho^+(x,t)\\ \rho^-(x,t)\end{bmatrix}. \tag{1.13}$$

If the particle is in the + state it is advected to the right (the $-U\partial_x$ term) or it jumps down to the – state at rate γ. In the – state the particle advects to the left or switches to the + state at rate γ.

The *marginal distribution* of the particle position is $\rho(x,t) = \rho^+(x,t)+\rho^-(x,t)$. Consider as well the auxiliary quantity $q(x,t) = \rho^+(x,t)-\rho^-(x,t)$. Adding and subtracting the differential equations in (1.13) we obtain evolution equations for ρ and for q:

$$(A) \qquad \partial_t\rho = -U\partial_x q \quad \text{and,}$$

$$(B) \qquad \partial_t q = -U\partial_x\rho - 2\gamma q.$$

If $q(x,0) = 0$ (meaning that $\rho^+ = \rho^-$ initially) the solution of (B) is

$$q(x,t) = -U\int_0^t e^{-2\gamma(t-s)}\,\partial_x\rho(x,s)\,ds = -U\partial_x\int_0^t e^{-2\gamma(t-s)}\,\rho(x,s)\,ds$$

and inserting this into (A) we find

$$\frac{\partial\rho}{\partial t} = U^2\partial_x^2\int_0^t e^{-2\gamma(t-s)}\,\rho(x,s)\,ds,$$

which is, in a certain sense, "close" to a diffusion equation.

Indeed, multiplying and dividing the expression above by 2γ we obtain

$$\frac{\partial\rho}{\partial t} = \frac{U^2}{2\gamma}\partial_x^2\int_0^t 2\gamma e^{-2\gamma(t-s)}\,\rho(x,s)\,ds,$$

where, on time scales larger than the velocity process' correlation time τ, the kernel mimics a delta distribution in s concentrated around t. More to the point, if we send $\gamma\to\infty$ and set $\frac{U^2}{2\gamma} = \kappa$, we are left with the diffusion equation

$$\frac{\partial\rho}{\partial t} = \kappa\frac{\partial^2\rho}{\partial x^2}.$$

Alternatively, we can directly compute the variance of the position process $X_t = \int_0^t UI_s\,ds$ finding

$$\mathbb{E}(X_t^2) = \frac{U^2}{2\gamma}\left(2t - \frac{1}{\gamma}(1-\exp(-2\gamma t))\right). \tag{1.14}$$

This calculation proceeds as follows. First note that

$$X_t^2 = \left(U \int_0^t I_s\, ds\right)\left(U \int_0^t I_r\, dr\right) = U^2 \int_0^t \int_0^t I_s I_r\, ds\, dr.$$

Then take the expectation to obtain

$$\begin{aligned}
\mathbb{E}(X(t)^2) &= U^2 \mathbb{E} \int_0^t \int_0^t I_s I_r\, ds\, dr \\
&= U^2 \int_0^t \int_0^t \mathbb{E}(I_s I_r)\, ds\, dr \\
&= U^2 \int_0^t \int_0^t \exp(-2\gamma|s - z|)\, ds\, dr.
\end{aligned}$$

Decompose the domain of integration as $\{0 \le s \le t, 0 \le z \le s\} \cup \{0 \le z \le t, 0 \le s \le z\}$ and observe that by the symmetry of the integrand

$$\begin{aligned}
\int_0^t \int_0^t \exp(-2\gamma|s - r|)\, ds\, dr &= \int_0^t \int_0^r \exp(-2\gamma(-s + r))\, ds\, dr \\
&\quad + \int_0^t \int_0^s \exp(-2\gamma(s - r))\, dr\, ds \\
&= 2 \int_0^t \int_0^r \exp(-2\gamma(-s + r))\, ds\, dr \\
&= 2 \frac{1}{2\gamma} \int_0^t \left(1 - \exp(-2\gamma r)\right)\, dr \\
&= 2 \frac{1}{2\gamma}\left(t + \frac{1}{2\gamma}(\exp(-2\gamma t) - 1)\right) \\
&= \frac{1}{2\gamma}\left(2t + \frac{1}{\gamma}(\exp(-2\gamma t) - 1)\right),
\end{aligned}$$

which coincides with (1.14).

Then it is clear that $E(X(t)^2) \sim \frac{U^2}{\gamma}(t - \tau)$ as $t \to \infty$ where $\tau = \frac{1}{2\gamma}$. Thus, for $t \gg \tau$ this random process disperses particles like a diffusion process (albeit with a short delay) with *effective diffusivity*

$$\kappa_{\text{eff}} := U^2 \frac{1}{2\gamma} = U^2 \tau.$$

The correlation time τ is the typical time it takes before changing velocity, a.k.a. the mean free time. It is also natural to define the mean free length $\ell = U\tau$ — the typical distance traveled before changing velocity — and write

$$\kappa_{\text{eff}} = U\ell.$$

This example shows how advection — in this case advection by a random-in-time velocity field — can disperse tracer particles diffusively, at least on sufficiently long time scales. In such systems the theoretical challenge is often to deduce an effective diffusivity in terms of details of the advecting flow field.

1.3 Can transport enhance diffusion?

Here we illustrate how a combination of advection and diffusion can enhance particle dispersion beyond the effect of either alone.

1.3.1 Shear flow

Consider a uniform shear flow $\boldsymbol{u} = \boldsymbol{i}Sy$ where $\boldsymbol{i}$ is the horizontal unit vector, S is the shear and y is the vertical component of the position vector $\boldsymbol{x} = \boldsymbol{i}x + \boldsymbol{j}y$.

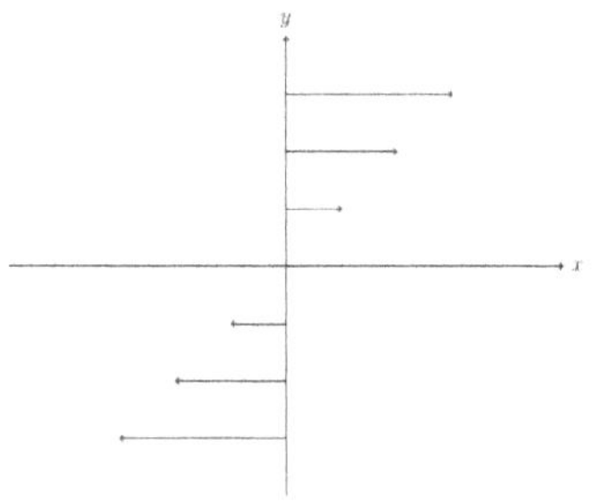

Fig. 1.4. Shear flow.

First consider motion in this flow field without diffusion. The horizontal and vertical positions of a passive tracer particle satisfy

$$dX_t = S\,Y_t\,dt,$$
$$dY_t = 0,$$

with initial data which we will take to be $X_0 = 0$ and $Y_0 = y_0$. Integrating the differential equations we find $X_t = Sy_0\,t$ and $Y_t = y_0$ from which we easily see the variance of the position to be $\mathbb{E}(X_t^2) = S^2\,Y_0^2\,t^2$. This sort of dispersion, $\mathbb{E}(X_t^2) \sim t^2$, is what we expect from ballistic motion.

Now include diffusion in the differential equations — making them truly stochastic differential equations — and consider

$$dX_t = SY_t\,dt + \sqrt{2\kappa}\,dW_t^{(1)},$$
$$dY_t = \sqrt{2\kappa}\,dW_t^{(2)},$$

where $W_t^{(1)}$ and $W_t^{(2)}$ are independent Wiener processes. Then the time-dependent joint probability density of particle's position $\rho(x, y, t)$ evolves according to the advection-diffusion partial differential (also known in various communities as a Fokker-Plank or Forward Kolmogorov) equation

$$\frac{d\rho}{dt} = -\frac{\partial}{\partial x}(Sy\rho) + \kappa\left(\frac{\partial^2}{\partial x^2} + \frac{\partial^2}{\partial y^2}\right)\rho,$$

a linear but non-constant coefficient partial differential equation.

We can easily compute the expectation values of X_t and Y_t. For simplicity suppose that $X_0 = 0 = Y_0$. Using the initial condition we have $Y_t = \sqrt{2\kappa}\,W_t$ and, since the expectation of the Wiener process is zero, $\mathbb{E}(Y_t) = 0$. From this it is easy to deduce that $\mathbb{E}(X_t) = 0$ as well.

The variance of X_t and Y_t follow from the Itô formula (1.7). Applied to Y_t^2 we see, as before, that

$$\mathbb{E}(Y_t^2) = 2\kappa t. \tag{1.15}$$

On the other hand applying the Itô formula to X_t^2 yields

$$\begin{aligned}\mathbb{E}(dX_t^2) &= 2\mathbb{E}(X_t(SY_t\,dt + \sqrt{2\kappa}dW_t)) + 2\kappa dt\\ &= 2S\mathbb{E}(X_tY_tdt) + 2\sqrt{2\kappa}\mathbb{E}(X_t\,dW_t) + 2\kappa dt\\ &= 2S\mathbb{E}(X_tY_t)dt + 2\kappa dt\,.\end{aligned} \tag{1.16}$$

It is left to compute $\mathbb{E}(X_tY_t)$. Using the Leibniz rule and (1.9) we find

$$\begin{aligned}\mathbb{E}(d(X_tY_t)) &= \mathbb{E}(Y_tdX_t + X_tdY_t + dX_tdY_t)\\ &= S\mathbb{E}(Y_t^2)dt + \sqrt{2\kappa}\mathbb{E}(Y_tdW_t) + \sqrt{2\kappa}\mathbb{E}(X_tdW_t)\\ &= 2S\kappa\,t\,dt.\end{aligned}$$

This implies that

$$\mathbb{E}(X_tY_t) = S\kappa t^2.$$

Inserting this into (1.16) we find

$$\mathbb{E}(dX_t^2) = d\,\mathbb{E}(X_t^2) = 2S^2\kappa t^2 dt + 2\kappa dt,$$

which upon integration yields

$$\mathbb{E}(X_t^2) = \frac{2}{3}S^2\kappa t^3 + 2\kappa t\,.$$

At early times molecular diffusion dominates particle dispersion, i.e., $\mathbb{E}(X_t^2) \sim t$, but later, in the presence of shear, the variance of the (horizontal) position grows $\sim t^3$. This super-diffusive motion is faster even than ballistic dispersion $\sim t^2$. Advection and diffusion together can disperse the particles faster than either alone!

1.3.2 Advection-diffusion equation

We have seen that mixing — thought of as passive tracer particle dispersion measured by position variance — can be modeled in two equivalent ways. For one, given an initial position, the stochastic trajectory of a passive tracer in $\mathbb{R}^d$ advected by a given incompressible flow field $\boldsymbol{u}$ (i.e., with $\nabla \cdot \boldsymbol{u} = 0$) may be described by the stochastic differential equation

$$d\mathbf{X}_t = \boldsymbol{u}(X_t, t)\, dt + \sqrt{2\kappa}\, d\mathbf{W}_t\,.$$

Equivalently a diffusing concentration field (density) field $T(\boldsymbol{x}, t)$ stirred by the incompressible fluid evolves according to the advection-diffusion equation

$$\dot{T} + \boldsymbol{u} \cdot \nabla T = \kappa \Delta T\,. \tag{1.17}$$

Previously we saw that if $\boldsymbol{u} = 0$ then the variance of the particle position is $\mathbb{E}(X_t^2) \sim 2\kappa t$. The density distribution of particles gets wider and wider. In this section we want to consider an alternative measure of mixing, namely the *suppression of the variance of the density.*

Consider the problem when the scalar field T satisfies the equation (1.17) in a bounded (and regular) domain $\Omega \subset \mathbb{R}^3$ with no-flux boundary conditions,

$$\boldsymbol{n} \cdot \boldsymbol{u} = 0 \quad \text{and} \quad \boldsymbol{n} \cdot \nabla T = 0 \quad \text{on } \partial\Omega, \tag{1.18}$$

where $\boldsymbol{n}$ is the outward normal to $\partial\Omega$.

Thanks to the incompressibility condition, we can rewrite (1.17) as

$$\dot{T} + \nabla \cdot \boldsymbol{J} = 0\,, \tag{1.19}$$

where $\boldsymbol{J} = \boldsymbol{u}T - \kappa\nabla T$. The field vector $\boldsymbol{J}$ has the units of a current,

$$[J] \approx \frac{\text{stuff}}{\text{area} \times \text{time}},$$

when the density T has units

$$[T] \approx \frac{\text{stuff}}{\text{volume}}$$

and the molecular diffusivity has units

$$[\kappa] = \frac{\text{length}^2}{\text{time}}\,.$$

If we isolate a small surface area in the domain and identify its normal unit vector $\boldsymbol{n}$, then $\boldsymbol{J}\cdot\boldsymbol{n}$ is the flux, the amount of stuff per unit time and per unit area, passing through that surface. (Similar considerations apply in lower dimensions: in $\mathbb{R}^2$ volumes are replaced by areas and area elements by line elements while in $\mathbb{R}^1$ "volumes" are line segments and "area elements" are their end points.)

Define the spatial average

$$\langle\cdot\rangle_{\boldsymbol{x}} = \frac{1}{|\Omega|}\int_{\Omega}(\cdot)d\boldsymbol{x}$$

and, for the purposes of this discussion, consider $\Omega \subset \mathbb{R}^n$ to be a spatially periodic domain $[0, 2\pi)^n$. Since there is no flux at the boundary – there are no boundaries – (1.17) implies

$$\frac{d}{dt}\langle T(\cdot, t)\rangle_{\boldsymbol{x}} = 0,$$

so that $\langle T(\cdot, t)\rangle_{\boldsymbol{x}}$ is constant in time. Then denote $\langle T\rangle_{\boldsymbol{x}} := \langle T(\cdot, t)\rangle_{\boldsymbol{x}}$ and decompose T into the stationary spatial mean $\langle T\rangle_{\boldsymbol{x}}$ and the deviation θ such that $T(\boldsymbol{x}, t) = \langle T\rangle_{\boldsymbol{x}} + \theta(\boldsymbol{x}, t)$. The deviation satisfies the same advection-diffusion equation

$$\dot{\theta} + \boldsymbol{u}\cdot\nabla\theta = \kappa\Delta\theta, \tag{1.20}$$

albeit with spatial mean zero at all times, $\langle\theta(\cdot, t)\rangle_{\boldsymbol{x}} = 0$.

When the density T is constant in space, i.e., when tracer particles are uniformly distributed and hence perfectly well mixed, the deviation $\theta = 0$. On the other hand when the solution is *not* well mixed then T is *not* constant and $\theta \neq 0$.

Therefore the spatial variance of T, i.e., $\langle(T(\cdot, t) - \langle T\rangle_{\boldsymbol{x}})^2\rangle_{\boldsymbol{x}} = \langle\theta(\cdot, t)^2\rangle_{\boldsymbol{x}}$ is a meaningful measure of the degree of "mixedness" of the tracers throughout Ω. The smaller the value of $\langle\theta^2\rangle_{\boldsymbol{x}}$ is, the better mixed the tracers are.

Testing the advection-diffusion equation (1.20) with θ and integrating by parts using the no-flux condition at the boundary, we obtain an expression for the evolution of the concentration variance

$$\frac{d}{dt}\langle\theta(\cdot, t)^2\rangle_{\boldsymbol{x}} = -2\kappa\langle|\nabla\theta|^2\rangle_{\boldsymbol{x}}. \tag{1.21}$$

Note specifically two things:

- molecular diffusion is necessary to make the concentration variance change (at least for sufficiently smooth densities such that $\nabla T = \nabla\theta$ is spatially square integrable), and
- the stirring flow field $\boldsymbol{u}$ does not explicitly appear in the variance evolution equation (1.21).

Suppose that $\boldsymbol{u} = 0$ and choose the initial data $\theta_0(\boldsymbol{x}) = A\sin(\boldsymbol{k}\cdot\boldsymbol{x})$ where $\boldsymbol{k}$ is a vector with integer components. Then the solution of (1.20) is $\theta(\boldsymbol{x}, t) =$

$A\exp(-\kappa k^2 t)\sin(\boldsymbol{k}\cdot\boldsymbol{x})$ where $k = |\boldsymbol{k}|$. Using the fact that $\langle \sin(\boldsymbol{k}\cdot\boldsymbol{x})^2\rangle_{\boldsymbol{x}} = \frac{1}{2}$, we see that the variance decays exponentially,

$$\frac{\langle\theta(\cdot,t)^2\rangle_{\boldsymbol{x}}}{\langle\theta_0^2\rangle_{\boldsymbol{x}}} = \exp(-2\kappa k^2 t), \tag{1.22}$$

with decay rate proportional to the diffusivity and inversely proportional to (the square of) the length scale in the density deviation. The question is, what role might stirring, i.e., advection with $\boldsymbol{u} \neq 0$, play in the variance suppression game? In fact stirring may greatly enhance the variance dissipating power of molecular diffusion to amplify mixing.

We illustrate this with a simple example. Focus for the moment on $\Omega = \mathbb{R}^2$ and consider a uniform shear flow

$$\boldsymbol{u} = \boldsymbol{i}Sy.$$

Then the advection-diffusion equation (1.20) is

$$\dot{\theta} + Sy\partial_x\theta = \kappa(\partial_x^2 + \partial_y^2)\theta\,, \tag{1.23}$$

and, when $\theta(x,y,0) = A\sin kx$, the solution is (exercise: check this)

$$\theta(x,y,t) = A\exp\left(-\kappa k^2(t + \frac{1}{3}S^2t^3)\right)\sin(k(x - Sty)).$$

The solution shows that the spatial structure of the density deviation remains sinusoidal but with a time-dependent wave-vector, $(\boldsymbol{i}k) \to (\boldsymbol{i}k - \boldsymbol{j}kSt)$. The shear flow transfers density fluctuations from the initial length scale $\sim k^{-1}$ to increasingly smaller scales $\sim \left(k\times\sqrt{1+S^2t^2}\right)^{-1}$. The concentration variance is

$$\langle\theta(\cdot,t)^2\rangle_{\boldsymbol{x}} \approx \frac{A^2}{2}\exp(-2\kappa k^2(t + \frac{1}{3}S^2t^3)), \tag{1.24}$$

where we have used $\langle\sin^2(k(x - Sty))\rangle_{\boldsymbol{x}} \approx \frac{1}{2}$ in sufficiently large and suitably shaped volumes $\Omega \subset \mathbb{R}^2$.

Shear stirring greatly enhances the exponential decay of the variance compared to the effect of diffusion alone. The flow field accomplishes this by tilting level sets of the initial condition which were aligned with the y–axis, forcing the wavelength to become smaller and smaller which increases concentration gradients to amplify the effect of molecular diffusion and increase the rate of variance dissipation.

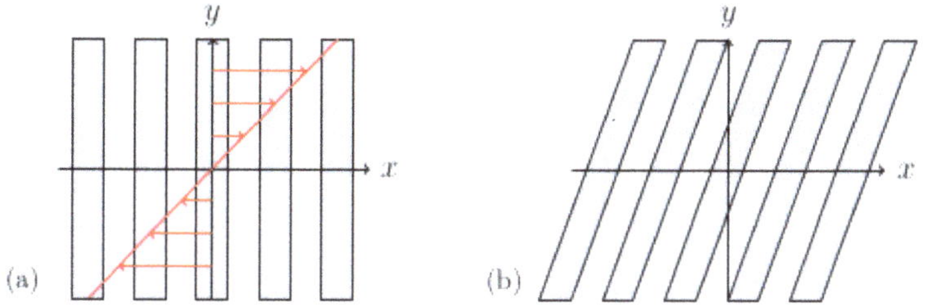

Fig. 1.5. The level set of the initial condition $\theta_0(x, y)$ are aligned with the y–axis and subsequently tilted by the shear flow.

Observe as well that if $\theta_0(x, y) = \sin ky$, i.e., if level sets of the concentration are initially aligned along the x axis, then the solution would be the same as for the diffusion-sans-advection equation. In that case this particular flow field plays *no* role in the concentration evolution. This example illustrates the importance of details of the alignment of density gradients and the flow — more precisely directions associated with rates of strain in the flow — in mixing processes as characterized by variance dissipation.

More generally, consider the evolution of the *concentration gradient*. The gradient of the advection-diffusion equation (1.20) is

$$\frac{d}{dt}\nabla\theta + \boldsymbol{u} \cdot \nabla\nabla\theta + \nabla\boldsymbol{u} \cdot \nabla\theta = \kappa\nabla\Delta\theta,$$

and, testing this with $\nabla\theta$ followed with integrations by parts, implies (when boundary terms vanish) that

$$\frac{d}{dt}\langle|\nabla\theta|^2\rangle_{\boldsymbol{x}} = -\kappa\langle|\Delta\theta|^2\rangle_{\boldsymbol{x}} - \langle\nabla\theta \cdot \nabla\mathbf{u} \cdot \nabla\theta\rangle_{\boldsymbol{x}}\,. \tag{1.25}$$

The first term on the right hand side above, the negative (semi)definite term, results from the effect of diffusion which strongly dissipates density fluctuations at small scales. The second term is the effect of stirring, showing how the symmetric part of $(\nabla\boldsymbol{u})$ — the *rate of strain* tensor $(\nabla\boldsymbol{u})_{\text{sym}}$ — affects the gradient variance.

For a two dimensional velocity field $\boldsymbol{u} = \boldsymbol{i}u + \boldsymbol{j}v$ the rate of strain tensor is

$$(\nabla\boldsymbol{u})_{\text{sym}} = \begin{bmatrix} u_x & \frac{v_x+u_y}{2} \\ \frac{v_x+u_y}{2} & v_y \end{bmatrix}.$$

At each point in space this matrix is symmetric and, thanks to the incompressibility condition $u_x + v_y = 0$, trace-free. Therefore whenever (generically) it is not identically zero, the rate of strain matrix has real and distinct eigenvalues $\lambda_\pm$ with orthogonal eigenvectors. The eigenvalues sum to zero so at generic non-degenerate points let $\lambda_+ >$

0 and $\lambda_- < 0$. The eigenvector associated with λ_+ is the *expansive* direction while the eigenvector associated with λ_- is the *contractive* direction.

In order for $(\nabla \boldsymbol{u})_{\text{sym}}$ to generate gradient variance (and thus contribute to concentration variance dissipation and increased mixing) $\nabla\theta$ should be predominantly aligned along the *contractive* directions. That is, level sets of θ should tend to be locally aligned with the *contractive* directions. Then $-\langle \nabla\theta \cdot \nabla \boldsymbol{u} \cdot \nabla\theta \rangle$ on the right hand side of (1.25) is positive which adds positively to $\frac{d}{dt}\langle |\nabla\theta|^2 \rangle$.

For the shear flow $\boldsymbol{u} = \boldsymbol{i}Sy$

$$\nabla \boldsymbol{u} = \begin{bmatrix} 0 & 0 \\ S & 0 \end{bmatrix},$$

and the rate of strain tensor is

$$(\nabla \boldsymbol{u})_{\text{sym}} = \frac{S}{2} \begin{bmatrix} 0 & 1 \\ 1 & 0 \end{bmatrix}.$$

The eigenvectors are $\begin{bmatrix} 1 \\ 1 \end{bmatrix}$ and $\begin{bmatrix} 1 \\ -1 \end{bmatrix}$ corresponding, respectively, to the maximally expanding and contractive directions. This steady shear flow is not an ideal gradient amplifier, however, because it changes the direction of the concentration gradients: sometimes it aligns concentration gradients along the contractive direction, but often the flow reorients them away as well.

1.3.3 Uniform strain flow

The *pure strain* flow, a divergence-free vector field of the form $\boldsymbol{u} = \boldsymbol{i}\Gamma x - \boldsymbol{j}\Gamma y$ with $\Gamma > 0$, has uniform gradient amplification properties. The trajectories of tracer particles are hyperbole and the rate of strain tensor is

$$(\nabla \boldsymbol{u})_{\text{sym}} = \Gamma \begin{bmatrix} 1 & 0 \\ 0 & -1 \end{bmatrix}.$$

The origin is a saddle point of the flow, the contractive direction is the y–direction and the expansive direction coincides with the x–direction.

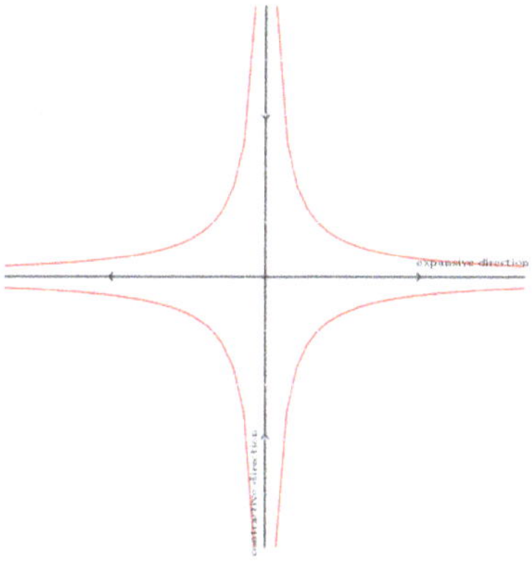

Fig. 1.6. The trajectories of the particles in the pure strain flow are hyperbole.

Maximal gradient amplification — and consequently maximal variance dissipation — is realized by initial data of the form $\theta_0(x, y) = A \sin k_0 y$ with level sets of the concentration fluctuations aligned with the expansive directions. With the pure strain flow in (1.20) the advection-diffusion equation becomes

$$\dot{\theta} + \Gamma x \frac{\partial}{\partial x}\theta - \Gamma y \frac{\partial}{\partial y}\theta = \kappa \left(\frac{\partial^2}{\partial x^2} + \frac{\partial^2}{\partial y^2} \right) \theta. \tag{1.26}$$

To solve it we make the ansatz $\theta(x, y, t) = \Theta(t) \sin(k(t)y)$ where we highlight the dependence of the wavenumber on time. Inserting the ansatz into the equation we find the exact solution (exercise: check this)

$$\theta(x, y, t) = A \exp\left(-\frac{\kappa k_0^2}{2\Gamma}(\exp(2\Gamma t) - 1) \right) \sin(k_0 \exp(\Gamma t) y) .$$

Tracer trajectories along the y–axis are squeezed exponentially into the origin and the wavelength $k(t)$ increases exponentially in time. This exponential amplification results in the double exponential decay of the amplitude $\Theta(t)$.

An alternative analysis proceeds by considering the system of two linear stochastic differential equations

$$dX_t = +\Gamma X_t dt + \sqrt{2\kappa} dW_t^{(1)}, \tag{1.27}$$

$$dY_t = -\Gamma Y_t dt + \sqrt{2\kappa} dW_t^{(2)}, \tag{1.28}$$

describing a particle moving and diffusing in the strain flow, where $W_t^{(1)}$ and $W_t^{(2)}$ are independent Wiener processes. The solutions X_t and Y_t are *Ornstein-Uhlenbeck processes* and the moments are easy to compute. Suppose initial conditions X_0 and Y_0. Then $\mathbb{E}(dX_t) = d\mathbb{E}(X_t) = \Gamma \mathbb{E}(X_t) dt$ and $\mathbb{E}(dY_t) = d\mathbb{E}(Y_t) = -\Gamma \mathbb{E}(Y_t) dt$ so

$$\mathbb{E}(X_t) = \mathbb{E}(X_0)\exp(\Gamma t) \quad \text{and} \quad \mathbb{E}(Y_t) = \mathbb{E}(Y_0)\exp(-\Gamma t) .$$

The Itô analysis (1.7-1.10) then implies

$$\mathbb{E}(dX_t^2) = d\mathbb{E}(X_t^2) = +2\Gamma\mathbb{E}(X_t^2)dt + 2\kappa dt,$$

$$\mathbb{E}(dY_t^2) = d\mathbb{E}(Y_t^2) = -2\Gamma\mathbb{E}(Y_t^2)dt + 2\kappa dt.$$

Thus if we specialize to $X_0 = 0 = Y_0$, then

$$\mathbb{E}(X_t^2) = \frac{\kappa}{\Gamma}(\exp(+2\Gamma t) - 1),$$

$$\mathbb{E}(Y_t^2) = \frac{\kappa}{\Gamma}(1 - \exp(-2\Gamma t)).$$

The variance of X_t grows exponentially while the variance of Y_t converges not to zero, but to $\frac{\kappa}{\Gamma}$ at $t \to \infty$. This means that a source point at the origin will be squeezed and stretched and eventually distributed over a thin layer of width $\left(\frac{\kappa}{\Gamma}\right)^{\frac{1}{2}}$ along the x−axis.

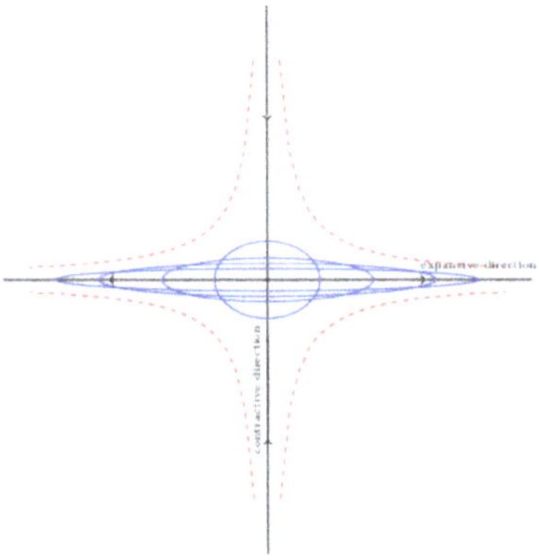

Fig. 1.7. A distribution of tracers is squeezed and stretched by the flow to be eventually distributed over a strip of vertical width $\left(\frac{\kappa}{\Gamma}\right)^{\frac{1}{2}}$.

1.3.4 The smallest scale of mixing

It is of great interest to understand the smallest dynamical scales in turbulent fluid flows. This remains a great challenge for turbulent solutions of the three dimensional Navier-Stokes equations but for other (simpler) problems the smallest scales can be identified. We ask, *what is the smallest length-scale we expect to see in passive tracer mixing?* In absence of diffusion “good” stirring creates very fine filaments which, due to incompressibility, become thinner and thinner as they are stretched longer and longer. In presence of diffusion, however, these filaments reach a minimal width; the filaments cannot get too thin because molecular diffusion broadens them. In the case of a pure strain flow the minimal width of the stretched blob is

$$\lambda_B = \left(\frac{\kappa}{\Gamma}\right)^{\frac{1}{2}}, \tag{1.29}$$

known as the *Batchelor scale*. The Batchelor scale quantifies the balance of strain (which stretches filaments and tends to make them thinner) with diffusion (which tends to broaden them).

For (homogeneous isotropic) turbulence we can relate the typical *strain rate* Γ with a quantity that characterizes the intensity of turbulence. Turbulence does not maintain itself; a source of energy must be supplied to sustain the motion and the relevant magnitude is the *energy dissipation rate per unit mass*

$$\epsilon := \nu\langle|\nabla u|^2\rangle_{\mathbf{x}}. \tag{1.30}$$

(Note: denoting the turbulent energy dissipation rate per unit mass by ϵ does *not* imply smallness in any sense; this is simply the traditional symbol for it in turbulence theory.) In turbulent flows at high Reynolds numbers – “small” viscosity – the flow stretches itself in order to increase its own velocity gradients which also enhances the effect of dissipation of energy at small scales due to viscosity. The energy dissipation rate ϵ is the power (per unit mass) required to maintain (statistical) steady state turbulence.

Identifying the typical rate of strain with the root mean square rate of strain, i.e., writing $\Gamma = \langle|\nabla u|^2\rangle_{\mathbf{x}}^{1/2}$, we have

$$\Gamma = (\epsilon\nu)^{\frac{1}{2}}.$$

Thus we can rewrite the Batchelor scale in terms of the energy dissipation rate

$$\lambda_B = \frac{\kappa^{\frac{1}{2}}}{\Gamma^{\frac{1}{2}}} = \frac{\kappa^{\frac{1}{2}}\nu^{\frac{1}{4}}}{\epsilon^{\frac{1}{4}}} = \left(\frac{\kappa}{\nu}\right)^{\frac{1}{2}}\left(\frac{\nu^3}{\epsilon}\right)^{\frac{1}{4}}.$$

Both ν and κ have the units of length^2/time, so $\frac{\kappa}{\nu}$ is nondimensional. The energy dissipation rate per unit mass ϵ has units length^2/time^3 and $(\nu^3/\epsilon)^{\frac{1}{4}}$ has units of a length. This quantity,

$$\lambda_k = \left(\frac{\nu^3}{\epsilon}\right)^{\frac{1}{4}},$$

is the *Kolmogorov scale*. It is supposedly the smallest length scale for the velocity field in turbulent flows.

The ratio $\frac{\nu}{\kappa}$ is called the *Prandtl number* (when κ is a thermal diffusion coefficient) or the *Schmidt number* (when κ is a mass diffusion rate). Prandtl numbers vary greatly according to the fluid under consideration. Liquid metals have Prandtl numbers $\ll 1$ while thick oils have Prandtl numbers $\gg 1$. Schmidt numbers are often high, though, and certainly so when the dissolved material has heavy molecular weight relative to the solvent. Thus in many applications the Batchelor scale $\lambda_B = \left(\frac{\kappa}{\nu}\right)^{\frac{1}{2}} \lambda_k$ is much smaller than the Kolmogorov scale. This means that one can have efficient and effective stirring even in mildly turbulent flows—and even at very small Reynold numbers.

1.3.5 Quantitative indicators of mixing efficacy

We now consider quantification of the mixing enhancement of stirring and examine how such measures might depend on the strength of the stirring.

Suppose the goal is to reduce the scalar variance from its initial value as much as possible in a certain time interval and reconsider the initial data $\theta_0(\boldsymbol{x}) = A\sin(\boldsymbol{k}_0 \cdot \boldsymbol{x})$. In absence of flow, i.e., $\boldsymbol{u} = 0$, the solution of (1.20) is $\theta(\boldsymbol{x}, t) = A\exp(-\kappa k_0^2 t)\sin(\boldsymbol{k}_0 \cdot \boldsymbol{x})$ where $k_0 = |\boldsymbol{k}_0|$ and the variance decays exponentially,

$$\frac{\langle \theta(\cdot, t)^2 \rangle_{\boldsymbol{x}}}{\langle \theta(\cdot, 0)^2 \rangle_{\boldsymbol{x}}} = \exp(-2\kappa k_0^2 t).$$

We can turn this formula inside out and note that molecular diffusivity may be expressed in terms of the time t_ε that it takes to reduce the variance to ε times its initial value:

$$\kappa = \frac{\ln \frac{1}{\varepsilon}}{2 t_\varepsilon k_0^2}. \tag{1.31}$$

In the presence of stirring we may still use (1.31) to define an *effective diffusion coefficient* in terms of the time t_ε that it takes to reduce the variance to ε times its initial value. That is, we gauge the effect of stirring by identifying the value of molecular diffusion that would be required to achieve the same level of variance suppression in time t_ε:

$$\kappa_{\text{eff}} = \frac{\ln \frac{1}{\varepsilon}}{2 t_\varepsilon k_0^2}. \tag{1.32}$$

In dimensionless terms we may consider *enhancement factor*

$$E := \frac{\kappa_{\text{eff}}}{\kappa}. \tag{1.33}$$

The task is to understand the dependence (i.e., scaling) of the dimensionless enhancement factor E on a dimensionless measure of the stirring strength commonly chosen to be the *Péclet number*

$$\text{Pe} = \frac{U\ell}{\kappa}, \tag{1.34}$$

where U is a relevant stirring speed, e.g, the root mean square stirring speed $U = \langle\langle|\boldsymbol{u}|^2\rangle_{\boldsymbol{x}}\rangle_t^{\frac{1}{2}}$, and ℓ is a suitable length scale.

We have seen that the variance of the concentration in the case of the shear flow decays super-exponentially, i.e., when $\boldsymbol{k}_0 = \boldsymbol{i}k_0$

$$\frac{\langle \theta(\cdot, t)^2 \rangle_{\boldsymbol{x}}}{\langle \theta(\cdot, 0)^2 \rangle_{\boldsymbol{x}}} = \exp(-2\kappa k_0^2 t(1 + S^2 t^2)).$$

Focusing on relatively long times ($St \gg 1$) we set $\varepsilon = \exp(-2\kappa k_0^2 S^2 t_\varepsilon^3) \ll 1$ so that

$$t_\varepsilon = \frac{(\ln \varepsilon)^{\frac{1}{3}}}{(2\kappa k_0^2 S^2)^{\frac{1}{3}}}.$$

Then by (1.33)

$$\kappa_{\text{eff}} = \frac{(2S \ln \frac{1}{\epsilon})^{\frac{2}{3}} \kappa^{\frac{1}{3}}}{k_0^{\frac{4}{3}}} \quad \text{and} \quad E = \frac{\kappa_{\text{eff}}}{\kappa} = \left(\frac{2S \ln \frac{1}{\epsilon}}{k_0^2 \kappa} \right)^{\frac{2}{3}}.$$

In the case of pure shear flow there is no natural length scale in the velocity field — it is entirely characterized by the shear strength S — so we must utilize the length scale in the initial concentration perturbation ($\ell \sim k_0^{-1}$) and use $U = \frac{S}{k_0}$ to identify the Péclet number $\text{Pe} = \frac{U\ell}{\kappa} = \frac{S}{k_0^2 \kappa}$. Then we observe

$$E = \frac{\kappa_{\text{eff}}}{k} \sim \text{Pe}^{\frac{2}{3}} \quad \text{as} \quad \text{Pe} \to \infty.$$

In this case, since $2/3 < 1$, $\kappa_{\text{eff}} = \kappa E \to 0$ as $\kappa \to 0$. That is, there is no residual mixing (by this measure) in the limit of zero molecular diffusivity.

We remark that if there is a definition of effective diffusivity κ_{eff} and a stirring such that the enhancement factor $E \sim \text{Pe}$, then $\kappa_{\text{eff}} \sim U\ell$ remains non-zero even in the singular limit of zero molecular diffusivity (all other parameters held fixed). In such situations κ_{eff} would be independent of the molecular diffusivity in the $\kappa \to 0$ limit. Then stirring would produce diffusive effects even if molecular diffusion is ostensibly negligible.

The pure strain flow, characterized by a constant rate of strain Γ, dissipates scalar variance faster than shear flow. Indeed, for pure strain flow and $\boldsymbol{k}_0 = \boldsymbol{j} k_0$ the effective diffusion definition in (1.33) yields $E = \frac{\kappa_{\text{eff}}}{\kappa} \sim \frac{\text{Pe}}{\ln \text{Pe}}$. To see this recall that the variance for the concentration associated to the pure strain flow is

$$\frac{\langle \theta(\cdot, t)^2 \rangle_{\boldsymbol{x}}}{\langle \theta(\cdot, 0)^2 \rangle_{\boldsymbol{x}}} = \exp\left(-\frac{\kappa k_0^2}{\Gamma} (\exp(2\Gamma t - 1)) \right).$$

Again focusing on relatively long times ($\Gamma t \gg 1$) we set $\varepsilon = \exp\left(-\kappa k_0^2 \exp(2\Gamma t_\varepsilon)/\Gamma\right)$ so that $t_\varepsilon = \frac{1}{2\Gamma} \ln\left(\frac{\ln(\frac{1}{\varepsilon}) 2\Gamma}{\kappa k_0^2}\right)$ and the effective diffusion definition (1.33) yields

$$\kappa_{\text{eff}} = \frac{\ln(\frac{1}{\varepsilon}) \Gamma}{k_0^2 \ln\left(\frac{\ln(\frac{1}{\varepsilon}) 2\Gamma}{\kappa k_0^2}\right)}.$$

Then with $\ell = \frac{1}{k_0}$ and $U = \frac{\Gamma}{k_0}$ so that $\text{Pe} = \frac{\Gamma}{k_0^2 \kappa}$ we obtain

$$E = \frac{\kappa_{\text{eff}}}{\kappa} \sim \frac{\text{Pe}}{\ln \text{Pe}}.$$

This is only logarithmically less that the $E \sim \text{Pe}$ scaling necessary to produce finite non-zero residual effective diffusion in the limit of vanishing molecular diffusion.

Later in these lectures we will see that there are definitions of effective diffusivity and stirring flow fields for which $E \sim \mathrm{Pe}^2$, meaning that effective diffusion can actually *increase* as molecular diffusivity decreases!

1.4 Sources and sinks

1.4.1 Internal source-sink distributions

Suppose now that Ω is a regular bounded domain of $\mathbb{R}^n$ and the velocity field $\boldsymbol{u}$ satisfies

$$\nabla \cdot \boldsymbol{u} = 0 \qquad \text{and} \qquad \boldsymbol{n} \cdot \boldsymbol{u} = 0 \text{ on } \partial\Omega,$$

where $\boldsymbol{n}$ is the outward normal to the domain. Let the scalar field $T(\boldsymbol{x}, t)$ satisfy

$$\begin{cases} \dot{T} + \boldsymbol{u} \cdot \nabla T &= \kappa \Delta T + S \quad \text{in} \quad \Omega \times (0, \infty)\,, \\ \boldsymbol{n} \cdot \nabla T &= 0 \quad \text{in} \quad \partial\Omega \times (0, \infty)\,, \\ T &= 0 \quad \text{in} \quad \Omega \times \{t = 0\}\,, \end{cases} \tag{1.35}$$

where $S = S(\boldsymbol{x})$ is a steady scalar source distribution.

Averaging in space using the no-flux boundary conditions we have

$$\frac{d}{dt}\langle T(\cdot, t)\rangle_{\boldsymbol{x}} = \langle S\rangle_{\boldsymbol{x}},$$

so that $\langle T(\cdot, t)\rangle_{\boldsymbol{x}} = \langle S\rangle_{\boldsymbol{x}} t$. Therefore it is natural to consider the decomposition

$$T(\boldsymbol{x}, t) = \langle S\rangle_{\boldsymbol{x}} t + \theta(\boldsymbol{x}, t),$$

where the deviation from the spatial average distribution satisfies

$$\dot{\theta} + \boldsymbol{u} \cdot \nabla\theta = \kappa\Delta\theta + s(\boldsymbol{x}), \tag{1.36}$$

with $s(\boldsymbol{x}) = S(\boldsymbol{x}) - \langle S\rangle_{\boldsymbol{x}}$. Therefore $\langle s\rangle_{\boldsymbol{x}} = 0$, $\theta(\boldsymbol{x}, 0) = 0$ and, subsequently, $\langle \theta(\cdot, t)\rangle_{\boldsymbol{x}} = 0$.

The efficiency of stirring to suppress the variance of the scalar concentration in this case is related to its ability to minimize θ (in the mean square sense). The inhomogeneous term $s(\boldsymbol{x})$ in (1.36) is signed – it is source in some places and a sink in others – and effective flows will transport tracers from source regions where $s(\boldsymbol{x}) > 0$ to sink regions where $s(\boldsymbol{x}) < 0$.

When $\boldsymbol{u} = 0$, (1.36) reduces to the inhomogeneous heat equation

$$\dot{\theta} = \kappa\Delta\theta + s(\boldsymbol{x}),$$

with steady state $\bar{\theta}(\boldsymbol{x}) = -\frac{1}{\kappa}(\Delta^{-1}s)(\boldsymbol{x})$ where $\Delta^{-1} = \nabla^{-2}$ denotes the appropriate inverse Laplacian operator. The variance of the scalar concentration in the "unstirred" steady state is

$$\langle \bar{\theta}^2\rangle_{\boldsymbol{x}} = \frac{\langle (\Delta^{-1}s)^2\rangle_{\boldsymbol{x}}}{\kappa^2}.$$

When the flow is non zero, i.e., $\boldsymbol{u} \neq 0$, the variance of the passive tracer concentration is

$$\langle\langle\theta^2\rangle_x\rangle_t = \frac{\langle\langle\theta^2\rangle_x\rangle_t}{\langle\bar{\theta}^2\rangle_x}\langle\bar{\theta}^2\rangle_x = \frac{\langle(\Delta^{-1}s)^2\rangle_x}{\kappa_{\text{eff}}^2}, \tag{1.37}$$

where $\langle\cdot\rangle_t$ is the time average and we have defined an effective diffusion coefficient

$$\kappa_{\text{eff}} := \sqrt{\frac{\langle(\Delta^{-1}s)^2\rangle_x}{\langle\langle\theta^2\rangle_x\rangle_t}} = \kappa_{\text{mol}}E,$$

and the *enhancement factor*

$$E := \sqrt{\frac{\langle\bar{\theta}\rangle_x}{\langle\langle\theta^2\rangle_x\rangle_t}}.$$

The effective diffusion coefficient–and its dimensionless version in terms of the enhancement factor–is a quantitative gauge of how much the stirring contributes to mixing as measured by concentration variance suppression. A basic challenge in the mixing business is to understand, predict and/or control this enhancement in terms of properties of the stirring velocity vector field. Another basic question is to determine the extent to which such effective diffusions or enhancement factors are meaningful as properties of a given flow field. That is, the degree to which κ_{eff} and E are properties of $\boldsymbol{u}$ alone–or how much they depend on the details of the task that the field is asked to perform e.g. the source-sink distribution $s(\boldsymbol{x})$, is not *a priori* evident.

1.4.2 Source-sink distribution at the boundary

Consider a rectangular box $\Omega = [0, L] \times [0, h]$ and, as before, suppose that $\boldsymbol{u}$ is given and satisfies

$$\nabla \cdot \boldsymbol{u} = 0 \qquad \text{and} \qquad \boldsymbol{n} \cdot \boldsymbol{u} = 0 \text{ in } \partial\Omega$$

and that the scalar field $T(\boldsymbol{x}, t)$ satisfies the homogeneous advection-diffusion equation with inhomogeneous boundary conditions:

$$\begin{cases} \dot{T} + \boldsymbol{u} \cdot \nabla T = \kappa \Delta T & \text{in } [0, L] \times [0, h] \times (0, \infty), \\ \nabla T \cdot \boldsymbol{n} = 0 & \text{at } y \in \{0, h\}, \\ T = T_0 + \delta T & \text{at } x = 0, \\ T = T_0 & \text{at } x = L. \end{cases} \tag{1.38}$$

In the absence of fluid flow, i.e., when $\boldsymbol{u} = \boldsymbol{i}u + \boldsymbol{j}v = 0$, then the steady solution is the linear profile $T_0 + \frac{\delta T}{L}(L - x)$ and the flux of the scalar, a.k.a. the current $\boldsymbol{J} \equiv \boldsymbol{i}J_x + \boldsymbol{j}J_y$, is $\boldsymbol{J} = \boldsymbol{i}J_x = \boldsymbol{i}\kappa G$ where the imposed gradient is $G = \frac{\delta T}{L}$. The linear relation between the current and the imposed gradient is known as Fick's law, and the coefficient of proportionality between the flux and the gradient is precisely the diffusion coefficient.

When the flow is non-zero then the local instantaneous current is

$$\boldsymbol{J}(\boldsymbol{x}, t) = \boldsymbol{i}J_x + \boldsymbol{j}J_y = \boldsymbol{u}(\boldsymbol{x}, t)T(\boldsymbol{x}, t) - \kappa\nabla T$$

and the space-time average of J_x is

$$\langle\langle J_x\rangle_{\boldsymbol{x}}\rangle_t = \langle\langle uT - \kappa\partial_x T\rangle_{\boldsymbol{x}}\rangle_t = \langle\langle uT\rangle_{\boldsymbol{x}}\rangle_t + \kappa G\,, \tag{1.39}$$

where in the last step we used the boundary conditions for T. The term $\langle\langle uT\rangle_{\boldsymbol{x}}\rangle_t$ is the correlation between the horizontal velocity and T.

Again we decompose the function $T(\boldsymbol{x}, t)$ into the steady no-flow state and deviation function θ:

$$T(\boldsymbol{x}, t) = T_0 + \frac{\delta T}{L}(L - x) + \theta(\boldsymbol{x}, t)\,.$$

The deviation satisfies homogeneous boundary conditions, $\theta = 0$ at $x = 0$ and $x = L$, with a space and time dependent source-sink distribution proportional to the horizontal velocity u:

$$\begin{cases} \dot{\theta} + \boldsymbol{u}\cdot\nabla\theta &= \kappa\Delta\theta + Gu \quad \text{in } [0, L]\times[0, h]\times(0, \infty)\,, \\ \theta &= 0 \quad \text{at } x \in \{0, L\}. \end{cases} \tag{1.40}$$

Note that for divergence-free $\boldsymbol{u}$ and the boundary conditions, $\langle u\phi\rangle_{\boldsymbol{x}} = 0$ for all the functions ϕ that depend only on x. Using this fact we find that the space-time average of J_x is

$$\langle\langle J_x\rangle_{\boldsymbol{x}}\rangle_t = \langle\langle u\theta\rangle_{\boldsymbol{x}}\rangle_t + \kappa G$$

and we can define the *effective diffusivity* as the ratio of the flux and the gradient:

$$\kappa_{\text{eff}} := \frac{\langle\langle J_x\rangle_{\boldsymbol{x}}\rangle_t}{G} = \frac{\langle\langle u\theta\rangle_{\boldsymbol{x}}\rangle_t + \kappa G}{G} = \kappa + \frac{\langle\langle u\theta\rangle_{\boldsymbol{x}}\rangle_t}{G}. \tag{1.41}$$

The *enhancement factor* in this context is

$$E := \frac{\kappa_{\text{eff}}}{\kappa} = 1 + \frac{\langle\langle u\theta\rangle_{\boldsymbol{x}}\rangle_t}{\kappa G}\,. \tag{1.42}$$

Testing the equation (1.40) with θ and using the incompressibility condition and the boundary conditions, we see that the evolution of the variance of θ is

$$\frac{d}{dt}\langle\theta(\cdot, t)^2\rangle_{\boldsymbol{x}} = -\kappa\langle|\nabla\theta|^2\rangle_{\boldsymbol{x}} + G\langle u\theta\rangle_{\boldsymbol{x}}\,.$$

By the maximum principle for the temperature T (and therefore also θ) is suitably bounded so that

$$\limsup_{t_0\to\infty} \frac{1}{t_0}\int_0^{t_0} \frac{d}{dt}\langle\theta(\cdot, t)^2\rangle_{\boldsymbol{x}}\,dt = \limsup_{t_0\to\infty} \frac{\langle\theta(\cdot, t_0)^2\rangle_{\boldsymbol{x}} - \langle\theta(\cdot, 0)^2\rangle_{\boldsymbol{x}}}{t_0} = 0.$$

Therefore $\kappa\langle\langle|\nabla\theta|^2\rangle_{\boldsymbol{x}}\rangle_t = G\langle\langle u\theta\rangle_{\boldsymbol{x}}\rangle_t$ and we can re-express the enhancement factor

$$E = \frac{\kappa_{\text{eff}}}{\kappa} = 1 + \frac{\langle\langle u\theta\rangle_{\boldsymbol{x}}\rangle_t}{\kappa G} = 1 + \frac{\langle\langle|\nabla\theta|^2\rangle_{\boldsymbol{x}}\rangle_t}{G^2}.$$

This tells us that the rate of transport of the scalar by the flow is proportional to $\langle\langle|\nabla\theta|^2\rangle_{\boldsymbol{x}}\rangle_t$, and in particular it ensures that $E \geq 1$. That is, flow of any sort can only *increase* the boundary-to-boundary scalar flux (on average).

1.5 Model stirring as diffusion

In this final section we develop a case study to address the following questions:

- How should we gauge the effectiveness of a flow as a mixer?
- How might we model advection and stirring as diffusion?
- What properties of a flow field make it a good mixer?

We saw that in order to investigate the mixing properties of a flow field we can consider the dispersion of tracer particles moving according to

$$dX_t = \mathbf{u}(X_t, t)\, dt + \sqrt{2\kappa}\, dW_t,$$

or we can study the concentration of the tracer particles and study solutions of

$$\partial_t T + \boldsymbol{u} \cdot \nabla T - \kappa \Delta T = S(\boldsymbol{x}),$$

where $S(\boldsymbol{x})$ is a scalar source distribution.

We might sensibly seek to quantify the mixing efficiency of stirring by replacing the flow field and diffusion with an effective diffusion that achieves the same goal. Symbolically, we seek to make sense of the substitution

$$\boldsymbol{u} \cdot \nabla - \kappa \Delta \to -\partial_i \kappa_{ij}^{\text{eff}} \partial_j,$$

and include the effect of stirring as an effective—usually *enhanced*—diffusion. (Not unexpectedly, a diffusion tensor would generally be necessary to model anisotropic flow features.) This approach raises several other fundamental questions:

- Which particular properties of “mixing” do we want to capture in the notion of an effective diffusion?
- Do different criteria for measuring mixing produce qualitatively different effective diffusivities?

Three reasonable ways to parametrize stirring as diffusion are (1) in terms of particle dispersion yielding an effective diffusion coefficient κ^{PD}, (2) as the ratio of a flux given a gradient in which case κ^{FG}, or (3) via concentration variance reduction with κ^{VR}. The second question above may be rephrased, for a given flow field are κ^{PD}, κ^{FG} and κ^{VR} the same? Or are they different?

In the first case the effective diffusivity is the time prefactor of the square-mean-displacement — at least if the time dependence is eventually linear:

$$\mathbb{E}\{(X_i(t) - X_i(0))(X_j(t) - X_j(0))\} \sim 2\kappa_{ij}^{\text{eff}}\, t.$$

In the second case we decompose the function T into a steady profile $T_0 - Gx$ and a deviation function $\theta = T - T_0 + Gx$ that satisfies

$$\dot{\theta} + \boldsymbol{u} \cdot \nabla\theta = \kappa\Delta\theta + G\,\boldsymbol{u} \cdot \boldsymbol{i}, \tag{1.43}$$

where $u = \boldsymbol{u} \cdot \boldsymbol{i}$ and $\boldsymbol{i}$ is the horizontal unit vector; recall (1.40). The source-sink here is proportional to the velocity in the direction of the gradient flux and the effective diffusivity is defined

$$\kappa_{11}^{\text{eff}} := \kappa + \frac{\langle\langle u\theta\rangle_{\boldsymbol{x}}\rangle_t}{G}\,. \tag{1.44}$$

As shown before, testing (1.43) with θ and integrating by parts implies this is

$$\kappa_{11}^{\text{eff}} = \kappa\left(1 + \frac{\langle\langle|\nabla\theta|^2\rangle_{\boldsymbol{x}}\rangle_t}{G^2}\right)\,. \tag{1.45}$$

In the third case we look at the deviation of the concentration variance in presence of a steady source-sink distribution, $\partial_t\theta + \boldsymbol{u} \cdot \nabla\theta = \kappa\Delta\theta + s(\boldsymbol{x})$, and define

$$\kappa^{\text{eff}} := \sqrt{\frac{\langle(\Delta^{-1}s)^2\rangle_{\boldsymbol{x}}}{\langle\langle\theta^2\rangle_{\boldsymbol{x}}\rangle_t}}\,. \tag{1.46}$$

This measure of effective diffusion expresses how much a flow reduces the scalar variance that is sustained by the inhomogeneous sources and sinks.

The Péclet number Pe is the non-dimensional parameter that measures the strength of the fluid flow relative to the molecular diffusion (see (1.34)). Large values of the Péclet number, i.e., $\text{Pe} \gg 1$, means that the stirring is strong while low values of Péclet, $\text{Pe} \ll 1$, indicate strong molecular diffusion and/or weak stirring. The goal of theory and analysis — and also often computation, simulation and/or experimentation — is to determine the enhancement factor as a function of the Péclet number. We are particularly interested in the limit $\kappa \to 0$ to see in which cases the effective diffusivity might become independent of the molecular diffusivity. But here a dilemma occurs. There are flows for which

$$E(\text{Pe}) = \frac{\kappa^{\text{PD}}}{\kappa} = \frac{\kappa^{\text{FG}}}{\kappa} \sim \text{Pe}^2 \quad \text{as} \quad \text{Pe} \to \infty,$$

while on the other hand [1] in terms of variance reduction

$$E(\text{Pe}) = \frac{\kappa^{\text{VR}}}{\kappa} \leq \text{Pe}^1 \quad \text{as} \quad \text{Pe} \to \infty.$$

The remainder of these lectures will demonstrate these facts and discuss resolution of the quandary.

To see the latter, consider a periodic box of size L with θ satisfying

$$\dot{\theta} + \boldsymbol{u} \cdot \nabla\theta = \kappa\Delta\theta + s(\boldsymbol{x}).$$

Multiply by a smooth test function $\phi(\boldsymbol{x})$ and average in $\boldsymbol{x}$ and t. Take the long time average in t (exercise: the first term vanishes by the suitable boundedness of θ) to obtain

$$\begin{aligned}\langle s\phi\rangle_{\boldsymbol{x}} &= -\langle\langle\theta(\boldsymbol{u}\cdot\nabla\phi+\kappa\Delta\phi)\rangle_{\boldsymbol{x}}\rangle_t \\ &\le \langle\langle\theta^2\rangle_{\boldsymbol{x}}\rangle_t^{\frac{1}{2}}\langle\langle(\boldsymbol{u}\cdot\nabla\phi+\kappa\Delta\phi)^2\rangle_{\boldsymbol{x}}\rangle_t^{\frac{1}{2}}\end{aligned}$$

using Cauchy-Schwarz in the last step. Then according to definition (1.46), for any test function ϕ

$$E=\frac{\kappa^{\mathrm{eff}}}{\kappa}\le\frac{1}{\kappa}\frac{\langle\langle(\Delta^{-1}s)^2\rangle_{\boldsymbol{x}}\rangle_t^{\frac{1}{2}}\langle\langle(\boldsymbol{u}\cdot\nabla\phi+\kappa\Delta\phi)^2\rangle_{\boldsymbol{x}}\rangle_t^{\frac{1}{2}}}{\langle s\phi\rangle_{\boldsymbol{x}}}. \tag{1.47}$$

To get the best upper bound, we would consider the infimum over all test functions

$$E\le\inf_{\phi}\left\{\frac{1}{\kappa}\frac{\langle\langle(\Delta^{-1}s)^2\rangle_{\boldsymbol{x}}\rangle_t^{\frac{1}{2}}\langle\langle(\boldsymbol{u}\cdot\nabla\phi+\kappa\Delta^{-1}\phi)^2\rangle_{\boldsymbol{x}}\rangle_t^{\frac{1}{2}}}{\langle s\phi\rangle_{\boldsymbol{x}}}\right\}, \tag{1.48}$$

but for our purposes it suffices to choose $\phi=\Delta^{-2}s$. (For some alternative analyses see [2].) When κ is small the dominant term on the right hand side is $\langle\langle(\boldsymbol{u}\cdot\nabla\phi\rangle_{\boldsymbol{x}}\rangle_t$ and, since $\frac{\nabla^{-3}s}{\Delta^{-1}s}$ has the unit of a length, we can employ it as ℓ to conclude that $E\lesssim\frac{U\ell}{\kappa}=\mathrm{Pe}^1$, where U is the root mean square stirring speed.

We reiterate that in this case the "suitable" length-scale ℓ employed in the definition of the Péclet number comes from the source-sink distribution function.

1.5.1 Single scale source-sink distribution

We now focus on the most fundamental example of a single scale source-sink distribution being stirred by a single-scale flow and explicitly compute some effective diffusions. Consider tracer particles diffusing in a steady sinusoidal shear flow satisfying the coupled stochastic (Itô) equations

$$\begin{aligned}dX_t &= \sqrt{2}U\sin(k_uY_t)\,dt+\sqrt{2\kappa}\,dW_t^{(1)},\\ dY_t &= \sqrt{2\kappa}\,dW_t^{(2)},\end{aligned}$$

where $W_t^{(1)}$ and $W_t^{(2)}$ are independent Wiener processes.

One thc onc hand, for a source-sink distribution $s(\boldsymbol{x})\sim\sin k_sx$, it was shown in [3] that the effective diffusion defined in terms of variance reduction according to (1.46) is actually $\kappa^{\mathrm{VR}}\sim\kappa\,(\frac{k_u}{k_s})^{1/3}\,\mathrm{Pe}^{5/6}$. (Here $\mathrm{Pe}=U/\kappa k_s$ using the length scale in the source-sink distribution.) This corresponds to an enhancement factor $E\lesssim\mathrm{Pe}^1$ in accord with the theorem.

On the other hand for the flux-gradient relation we need to compute the steady state solution of (1.43) with velocity field of the form $\boldsymbol{u}=\sqrt{2}U\sin(k_uy)\boldsymbol{i}$:

$$\partial_t\theta+\sqrt{2}U\sin(k_uy)\partial_x\theta=\kappa\Delta\theta+G\sqrt{2}U\sin(k_uy)\,.$$

The steady-state solution is

$$\theta_S(y) = \frac{\sqrt{2}Gu}{\kappa k_u^2} \sin(k_u y),$$

and the enhancement factor can be easily computed by the definition (see (1.45)):

$$E = \frac{\kappa^{FG}}{\kappa} = 1 + \frac{\langle\langle|\nabla\theta_S|^2\rangle_x\rangle_t}{G^2} = 1 + \mathrm{Pe}^2\,.$$

Moreover, we can compute the (enhanced) particle dispersion in the x direction exactly. Noting that

$$dX_t^2 = 2X_t dX_t + \frac{1}{2}2\,(dX_t)^2,$$

and $\mathbb{E}((dX_t)^2) = 2\,\kappa\,\mathbb{E}((dW_t^{(1)})^2) + o(dt) = 2\,\kappa\,dt$, taking the expectation value of the expression above, we have

$$\mathbb{E}(dX_t^2) = 2\mathbb{E}(X_t dX_t) + 2\kappa dt = 2\sqrt{2}\,U\,\mathbb{E}(X_t \sin(k_u Y_t))\,dt + 2\,\kappa\,dt.$$

In order to compute the term $\mathbb{E}(X_t \sin(k_u Y_t))$ we use again the Itô formula to compute

$$\begin{aligned} d(X_t \sin(k_u Y_t)) &= \sin(k_u Y_t)dX_t + X_t(k_u \cos(k_u Y_t))dY_t - \frac{1}{2}X_t k_u^2 \sin(k_u Y_t)2\kappa dt \\ &= \sin(k_u Y_t)(\sqrt{2}U \sin(k_u Y_t)\,dt + \sqrt{2\kappa}\,dW_t^{(1)}) \\ &\quad +X_t(k_u \cos(k_u Y_t))\sqrt{2\kappa}\,dW_t^{(2)} - \frac{1}{2}X_t k_u^2 \sin(k_u Y_t)2\kappa dt\,. \end{aligned}$$

Taking the expectation value of the last expression we have

$$\mathbb{E}(d(X_t \sin(k_u Y_t))) = \sqrt{2}U\mathbb{E}(\sin^2 k_u Y_t)dt - \kappa k_u^2 \mathbb{E}(X_t \sin(k_u Y_t))\,dt.$$

Since $\mathbb{E}(\sin^2(k_u Y_t)) = \frac{1}{2}$, we are left to solve

$$\mathbb{E}(d(X_t \sin(k_u Y_t))) = d\mathbb{E}(X_t \sin(k_u Y_t)) = (U/\sqrt{2})\,dt - \kappa k_u^2 \mathbb{E}(X_t \sin(k_u Y_t))\,dt.$$

The solution is

$$\mathbb{E}(X_t \sin(k_u Y_t)) = \exp(-\kappa k_u^2 t) + U/(\sqrt{2}\,\kappa\,k_u^2).$$

In the long-time limit $\mathbb{E}(X_t \sin(k_u Y_t)) \to \frac{U}{\sqrt{2}\kappa k_u^2}$, so the variance of X_t satisfies

$$\mathbb{E}(dX_t^2) = d\mathbb{E}(X_t^2) \to \frac{2U^2}{\kappa k_u^2}dt + \kappa dt$$

and

$$\mathbb{E}(X_t^2) \to 2\left(\frac{U^2}{\kappa k_u^2} + 2\kappa\right)t.$$

Thus $\kappa_{11}^{\text{eff}} = \kappa + \frac{U^2}{\kappa k_u^2}$, and using $\ell \sim k_u^{-1}$ as the reference length-scale so that $\mathrm{Pe} = \frac{U^2}{\kappa k_u^2}$, we have $E = \frac{\kappa_{11}^{\text{eff}}}{\kappa} = \frac{\kappa^{PD}}{\kappa} = 1 + \mathrm{Pe}^2$.

The origin of the different behavior of the different definitions of effective diffusion – i.e., $E \sim \mathrm{Pe}^1$ for variance reduction versus $E \sim \mathrm{Pe}^2$ for the flux-gradient and particle dispersion measures – resides in a combination of the different choices for the length scale ℓ and consideration of the different times scales over which the mixing processes take place.

Indeed, it takes time $\sim (\kappa k_u^2)^{-1}$ for tracers to develop an effective diffusivity $\sim \mathrm{Pe}^2$ since trajectories must diffuse across various shear "streams" before the flow-induced dispersion enhancement to emerge. If this time is as long or longer than the time it takes for the flow to transport tracer from sources to sinks, i.e., $(Uk_s)^{-1}$, then then Pe^2 enhancement is irrelevant for variance reduction. That is, $E = \mathcal{O}(\mathrm{Pe}^2)$ enhancement is only possible when $\mathrm{Pe} < \frac{k_u}{k_s}$ (where the Péclet number is defined with $\ell \sim k_u^{-1}$).

Ultimately the difference arises due to fundamental physical features of *transient* mixing, such as might be measured by tracer particle dispersion, and *transport* properties of a flow that are relevant to steady state variance suppression in the presence of sources and sinks. For further discussion see [4].

Acknowledgment: The authors would like to thank Anna Mazzucato and Gianluca Crippa for organizing the very interesting Summer School in Levico Terme, and the Fondazione Bruno Kessler und the CIRM for their administrative support and hospitality. Much of the research reported here was supported in part by US National Science Foundation Division grants including, most recently, PHY-1205219 and DMS-1515161. The first author (CRD) gratefully acknowledges support as a Simons Fellow in Theoretical Physics and as a Fellow of the John Simon Guggenheim Foundation during, respectively, the presentation of these lectures and preparation of these notes.

Appendix: Itô Formula

The Wiener process is a Gaussian distribution with $\mathbb{E}(W_t) = 0$, $\mathbb{E}(W_t^2) = t$ and $\mathbb{E}(W_t, W_s) = \min(t, s)$. Consider the independent increments

$$\Delta W_t = W_{t+\Delta t} - W_t,$$

and compute their expectation values:

$$\begin{aligned} \mathbb{E}(\Delta W_t^2) &- \mathbb{E}(W_{t+\Delta t} - W_t)^2 \\ &= \mathbb{E}(W_{t+\Delta t}^2 - 2W_{t+\Delta t}W_t + W_t^2) \\ &= t + \Delta t - 2t + t = \Delta t\,. \end{aligned} \tag{1.49}$$

Therefore,

$$\mathbb{E}(dW_t^2) = dt\,.$$

Now consider a stochastic process defined by

$$dX_t = f(X_t)dt + g(X_t)\,dW_t,$$

where W_t is a Wiener process, interpreted via the forward Euler scheme

$$\Delta X_t = X_{t+\Delta t} - X_t = f(X_t)\Delta t + g(X_t)\,\Delta W_t .$$

Note that ΔW_t is independent of X_t.

Let us compute the variance of the increment ΔX. We have

$$\begin{aligned}\mathbb{E}(\Delta X_t^2) &= \mathbb{E}(f(X_t)^2(\Delta t)^2) + 2\mathbb{E}(f(X_t)g(X_t)\Delta W_t \Delta t) + \mathbb{E}(g(X_t)^2 \Delta W_t^2)\\ &= \mathbb{E}(f(X_t)^2)\,(\Delta t)^2 + \mathbb{E}(g(X_t)^2)\,\Delta t\\ &\approx \mathbb{E}(g(X_t)^2)\,\Delta t\,. \end{aligned} \tag{1.50}$$

The Itô formula states that the differential of any function F of the process X_t is

$$dF(X_t) = F'(X_t)\,dX_t + \frac{1}{2}F''(X_t)\,(dX_t)^2, \tag{1.51}$$

keeping only the $\mathcal{O}(dt)$ terms, i.e., those in dX_t and using $(dX_t)^2 \approx g(X_t)^2\,dt$.

For example we use the Itô formula (1.51) applied to the function $F : X_t \to X_t^2$ in order to compute the variance of stochastic processes. Namely,

$$d(X_t^2) = 2X_t dX_t + \frac{1}{2}2(dX_t)^2.$$

Taking the expectation value of (1.7) we have

$$\mathbb{E}(dX_t^2) = 2\mathbb{E}(X_t dX_t) + \mathbb{E}(dX_t)^2.$$

Bibliography

[1] Jean-Luc Thiffeault, Charles R. Doering, and John D. Gibbon. A bound on mixing efficiency for the advection–diffusion equation. Journal of Fluid Mechanics, 521:105–114, 2004.

[2] Charles R. Doering and Jean-Luc Thiffeault. Multiscale mixing efficiencies for steady sources. Physical Review E, 74:025301R, 2006.

[3] Tiffany A. Shaw, Jean-Luc Thiffeault, and Charles R. Doering. Stirring up trouble: Multi-scale mixing measures for steady scalar sources. Physica D, 231:143–164, 2007.

[4] Zhi Lin, Katarína Bodová, and Charles R. Doering. Models and measures of mixing and effective diffusion. Discrete and Continuous Dynamical Systems, 28:259–274, 2010.

Yann Brenier* and Laura Gioia Andrea Keller

New concepts of solutions in fluid dynamics

Abstract: The aim of these lecture notes is to present some new possibilities of defining *generalized flows* and *generalized solutions to the Euler equations*.
Roughly speaking, the approach which we emphasize here arises from the idea of probability measures on paths and is motivated by the hydrostatic rescaling of the Euler equations.
We will start with a short introduction to fluid mechanics, including the Eulerian as well as the Lagrangian point of view, then pass on to the presentation of the notion of generalized flows and generalized solutions to the Euler equations and finally we will give some examples and connections to related topics.
These notes by far can not cover all the relevant material but we give suggestions for further reading and indications to related subjects and other approaches.

Keywords: Fluid dynamics, generalized flows, generalized solutions to Euler equations, probability measures, least action principle

2.1 Introduction

We will start from a brief review of the Eulerian as well as the Lagrangian viewpoint of describing fluid motions.
Starting from these, the main features which permit to give new notions of generalized solutions are discussed. These new notions are presented and discussed in the following sections.

Eulerian description of fluid motion and hydrostatic rescaling of the Euler equations

We start with the Eulerian description of a fluid flow, a description in terms of a velocity vector field in a fixed reference ("laboratory") frame.
We look at a domain of the form

$$\tilde{B} = D \times (0, \varepsilon),$$

where D is a 2-dimensional domain (e.g. a periodic square) and ε represents the height of the 3-dimensional "box" $\tilde{B}$.
A point in $\tilde{B}$ is described by $\tilde{x} = (\tilde{x}_1, \tilde{x}_2, \tilde{x}_3)$.

***Corresponding Author: Yann Brenier:** Centre de Mathématiques, Laurent Schwartz, Ecole Polytechnique, France, E-mail: yann.brenier@math.polytechnique.fr
Laura Gioia Andrea Keller: Lucerne University of Applied Sciences and Arts, Switzerland, E-mail: laura.keller@math.ethz.ch

Moreover, the pressure is denoted by $\tilde{p}(t, \tilde{x}) \in \mathbb{R}$ (t as usual stands for the time), the velocity is denoted by $\tilde{v}(t, \tilde{x}) = (\tilde{v}_1, \tilde{v}_2, \tilde{v}_3)$ and we assume an uniform density of the fluid at hand.
Then, the motion of the fluid is described by the Euler equations (equations (E)) together with the incompressibility constraint (equation (IC))

$$\begin{cases} \partial_t \tilde{v} + (\tilde{v} \cdot \tilde{\nabla})\tilde{v} + \tilde{\nabla}\tilde{p} = 0 \text{ on } \tilde{B} \times \mathbb{R}^+, & (E) \\ \tilde{\nabla} \cdot \tilde{v} = 0 \text{ on } \tilde{B} \times \mathbb{R}^+. & (IC) \end{cases}$$

This description in terms of partial differential equations (PDEs) is called the Eulerian description of fluid motion and can also be seen as a limit case of the Navier-Stokes equations, namely the case of vanishing kinematic viscosity $\nu = 0$.
This description was established by Euler in [24] and a more modern presentation and derivation of these equations can be found e.g. in the monograph of Landau and Lifshitz [26].

Of course the above equations alone do not give a complete problem. In addition, boundary and initial conditions have to be imposed. Here at this point, we will not enter into a particular choice of such data.
Since we want to focus on the hydrostatic limit, the boundary conditions we fix are the following ones

$$\tilde{v}_3 = 0 \quad \text{on} \quad \{\tilde{x}_3 = 0 \text{ or } \tilde{x}_3 = \varepsilon\}.$$

Then, in a first step, we will rescale the vertical coordinate. The new coordinates are

$$x_1 = \tilde{x}_1, \; x_2 = \tilde{x}_2 \quad \text{and} \quad x_3 = \varepsilon^{-1}\tilde{x}_3.$$

Thus, the new vertical coordinate takes values in $(0, 1)$ and the new domain is denoted by B, i.e., $B = D \times (0, 1)$.
Accordingly, v and p denote the velocity and the pressure in these new coordinates. More precisely, we have $p(x_1, x_2, x_3) = \tilde{p}(x_1, x_2, \varepsilon x_3)$, $v_1(x_1, x_2, x_3) = \tilde{v}_1(x_1, x_2, \varepsilon x_3)$ and $v_2(x_1, x_2, x_3) = \tilde{v}_2(x_1, x_2, \varepsilon x_3)$. In the case of v_3 an additional scaling is necessary in order to prevent the formation of shocks, namely $v_3(x_1, x_2, x_3) = \varepsilon^{-1}\tilde{v}_3(x_1, x_2, \varepsilon x_3)$.
Next, we want to see how the Euler equations transform to these new coordinates:
It is easy to check that still we have

$$\partial_1 v_1 + \partial_2 v_2 + \partial_3 v_3 = 0,$$

and

$$\tilde{v} \cdot \tilde{\nabla} = v \cdot \nabla = v_1\partial_1 + v_2\partial_2 + v_3\partial_3.$$

Then, we introduce the following "abbreviation" for the horizontal parts

$$x_H = (x_1, x_2), \quad \text{i.e.,} \quad x = (x_1, x_2, x_3) = (x_H, x_3),$$

and accordingly

$$v_H = v_H(t, x_H, x_3) = (v_1, v_2),$$

which finally leads to the Euler equations in the new coordinates

$$\begin{cases} \partial_t v_H + (v \cdot \nabla) v_H + \nabla_H p = 0 & \text{on } B \times \mathbb{R}^+, \\ \varepsilon^2 \left(\partial_t v_3 + (v \cdot \nabla) v_3\right) + \partial_3 p = 0 & \text{on } B \times \mathbb{R}^+. \end{cases}$$

If we let ε tend to zero we obtain the so called hydrostatic limit, the "hydrostatic Euler equations"

$$\begin{cases} \partial_t v_H + (v \cdot \nabla) v_H + \nabla_H p = 0 \text{ on } B \times \mathbb{R}^+, & (hE_H) \\ \partial_3 p = 0 \text{ on } B \times \mathbb{R}^+, & (p3) \\ \nabla \cdot v = 0 \text{ on } B \times \mathbb{R}^+. & (IC) \end{cases}$$

Remark 2.1. *The operator*

$$D_t = \partial_t + v \cdot \nabla$$

is called material derivative and using this abbreviation, the equation (hE_H) *can be rewritten as*

$$D_t v_H + \nabla_H p = 0.$$

In addition, the kinetic energy

$$\tilde{E}_{kin} = \frac{1}{2} \int_D \int_0^\varepsilon |\tilde{v}(t, \tilde{x})|^2 \, d\tilde{x} = \frac{1}{2} \int_D \int_0^\varepsilon \tilde{v}_1^2 + \tilde{v}_2^2 + \tilde{v}_3^2 \, d\tilde{x}$$

reads as follows in the new coordinates

$$E_{kin} = \frac{1}{2} \int_D \int_0^1 |v(t, x)|^2 \, dx = \frac{1}{2} \int_D \int_0^1 v_1^2 + v_2^2 + \varepsilon^2 v_3^2 \, dx,$$

and becomes thus, in the limit as ε *tends to zero,*

$$E = \frac{1}{2} \int_D \int_0^1 |v_H(t, x)|^2 \, dx.$$

This new energy E will be used later on when concepts through the principle of "least action" are discussed.

The interested reader is referred to the first author's article [13] for a more detailed discussion of the relation between the original Euler equations and the hydrostatic limit.

Lagrangian description of fluid motion

As a second point of view, we will here recall the Lagrangian description of a fluid flow. Roughly speaking, here the point of view is to "follow a fluid particle on its trajectory". In this case, the motion is governed by the following ordinary differential equation (ODE)

$$\frac{d}{dt}\xi(t,a) = v(t,\xi(t,a)), \tag{L1}$$

where $\xi(t,a)$ is the position at time t of the fluid particle with initial position

$$a = \xi(0,a)$$

in the physical 3-dimensional domain D^3, e.g. $D^3 = B$ from above. Sometimes, a is also called the Lagrangian particle marker.
In addition, v and p have the same meaning as before in the Eulerian description.

Remark 2.2. *The connection between the Eulerian and the Lagrangian description of the fluid motion is given by the theory of characteristics (see Evans' book [25] for a short introduction to the theory of characteristics). More precisely, the Lagrangian particle trajectories $\xi(t,\cdot)$ are characteristics of the material derivative D_t: For $v(t,x) = v(t,\xi(t,a))$ it holds*

$$\frac{d}{dt}v(t,\xi(t,a)) = \partial_t v + \nabla v \cdot \frac{d\xi}{dt} = v_t + v \cdot \nabla v.$$

Summarized, the relations between the two points of view are

$$v = \frac{d\xi}{dt} \quad \text{and} \quad x = \xi(t,a).$$

Back to equation (L1), the Euler equations $\partial_t v + (v \cdot \nabla)v + \nabla p = 0$ we have seen above then imply

$$\frac{d^2\xi}{dt^2} = -(\nabla p)(t,\xi(t)), \tag{L2}$$

meaning that the acceleration is given by the pressure.

If the involved velocity is smooth enough, the Cauchy-Lipschitz theory allows to solve the above ODE for ξ. This provides the most classical setting in which the problem can be studied from the Lagrangian point of view. A drawback is that this theory requires more regularity that what can be assumed in physically interesting situations. Thus, a more flexible theory is desired. Below, we will give further comments on this issue.

Another important feature of the Lagrangian flow equation - under the additional constraint of incompressibility - is that for all fixed times t, $\xi(t,\cdot)$ is a diffeomorphism from $D^3 \subset \mathbb{R}^3$ to $\xi(D^3,t) = \{\xi(t,a)|\ a \in D^3\}$ with Jacobian determinant

$$J_a = det\big(\nabla_a \xi(t,a)\big) = 1 \quad \text{for all times } t.$$

In what follows, we will mainly restrict ourselves to the case, where the physical domain D^3 is fixed, i.e., we look at a situation where the fluid moves inside D^3. The "compartment" in which the fluid is bound does not change but only the configuration of the fluid changes. Moreover, often we will limit ourself to the case when D^3 is the unit cube.

The above property for incompressible fluids can be rephrased alternatively in the following three equivalent ways - recall that we look at the situation where $\xi : D^3 \to D^3$ for all times t:

i) *Weak formulation of incompressibility*
   For every $t \geq 0$ and for every continuous function f it holds

$$\int_{D^3} f(\xi(t,a))\, da = \int_{D^3} f(a)\, da.$$

ii) *Measure theoretic formulation of incompressibility*
   For every Borel set $\mathcal{B} \subset D^3$ its Lebesgue measure $|\mathcal{B}|$ is preserved

$$|\xi(t,\cdot)^{-1}(\mathcal{B})| = |\mathcal{B}|.$$

   In other words, the push forward of the Lebesgue measure is again the Lebesgue measure, $\xi_\sharp|\cdot| = |\cdot|$.

iii) The image of $da|_{D^3}$ under the map $\xi(t,\cdot)$ is again $da|_{D^3}$.

A derivation of the equivalence of the above three formulations can be found e.g. in the monograph of Majda-Bertozzi [27].

Next, we will combine the observations made for the hydrostatic Euler equations and for the Lagrangian equation of motion, in particular with the second order version (L2).
To this end, we write in a first step $a = (a_H, a_3)$ and $\xi = (\xi_H, \xi_3)$ according to the new coordinates we introduced at the end of the first paragraph. Then, due to the weak formulation of the incompressibility condition we have for every continuous function f

$$\int_{D_H}\int_0^1 f(\xi_H(t,(a_H,a_3)), \xi_3(t,(a_H,a_3)))\, da_H da_3 = \int_{D_H}\int_0^1 f(a_H,a_3)\, da_H da_3.$$

In particular, if we restrict our attention to "horizontal" continuous functions $f \in C(D_H)$, i.e., $f = f(x_1, x_2)$, we have due to the above property

$$\begin{aligned}\int_{D_H}\int_0^1 f(\xi_H(t,(a_H,a_3)))\,da_H da_3 &= \int_{D_H}\int_0^1 f(a_H,a_3)\,da_H da_3\\ &= \int_{D_H} f(a_H)\,da_H. \end{aligned} \tag{2.1}$$

Next, we observe that we have

$$\frac{d^2\xi_H}{dt^2} = -\nabla_H p(t,\xi_H(t,a),\xi_3(t,a)) = -\nabla_H p(t,\xi_H(t,a)),$$

where the second equality holds due to the fact that $\partial_3 p = 0$, which implies that $\nabla_H p$ does not depend on ξ_3.

So all together we are left with a description of our flow in terms of

$$\xi_H(t,a_H,a_3), \quad \text{and} \quad p(t,x_H),$$

where

$$a_H, x_H \in D_H \quad \text{and} \quad a_3 \in [0,1].$$

Thus, $a_H = (x_1, x_2)$ and x_H are the new geometric and physical variables, whereas a_3 becomes a "microstructure" variable.

Note that once ξ_H is known, ξ_3 can be reconstructed from the condition

$$det\big(\nabla_a \xi(t,a)\big) = 1.$$

To summarize, through the hydrostatic rescaling and limit, we have obtained, in the limit, a purely horizontal model on the 2-dimensional domain D_H, where:

i) ξ_3 has dropped out as a geometric variable to become a "microstructure" variable,
ii) the governing pressure p is now purely 2-dimensional (since $\partial_3 p = 0$).

This hydrostatic limit will be for us the "model" of a generalized solution.

Remarks 2.3.

i) A more profound comparison between the Eulerian and the Lagrangian approach can be found in the book of Crippa [20], also in view of possible generalized concepts of flows and solutions.
ii) Sometimes, it is useful to combine the Eulerian and the Lagrangian description of fluid motion. An example of such an approach can be found in Constantin's articles [18] and [19].

2.2 A combinatorial-numerical illustration

As we have seen before, the “prototype” of a flow we want to look at and which should give us the right starting point in order to generalize the concept of flow, was a “horizontal flow” ξ_H with some additional “microstructure” a.

This seems to be a violation of the fundamental physical principle that at each point in space there can be only one trajectory passing through this point (see also the first author’s slides [5]).
The aim of this section is to present a discrete toy model which illustrates this behavior.

In order to understand this better, we look at the following example.
We start with the following arrangement of four elements, labelled for simplicity by 1 to 4.

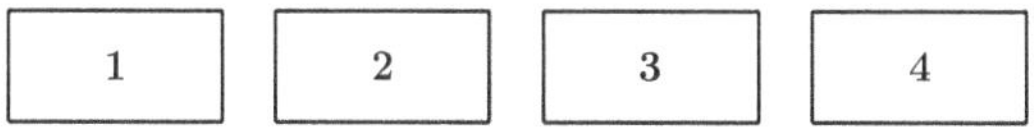

Fig. 2.1. Initial arrangement

Then our goal it to rearrange these elements such that thy appear in reversed order

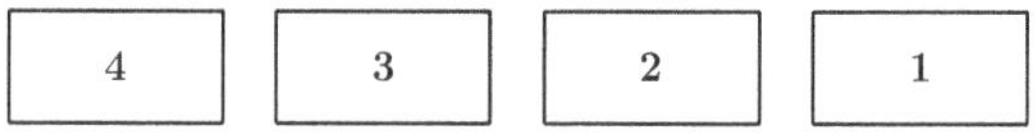

Fig. 2.2. Final arrangement

In fact, this end can be reached by simple permutations, where in each step neighboring elements are flipped. This procedure is explained in the figure below.
If we draw also the trajectories we see that the trajectories intersect (see below). A behavior which physically is forbidden! But if we “lift” the problem with an extra “hidden hight” variable we can get rid of the problem of intersecting trajectories!

The moral of this approach is that due to an additional dimension - the introduction of an additional “microstructure” - we can handle a situation which in the original dimension seems non-physical.

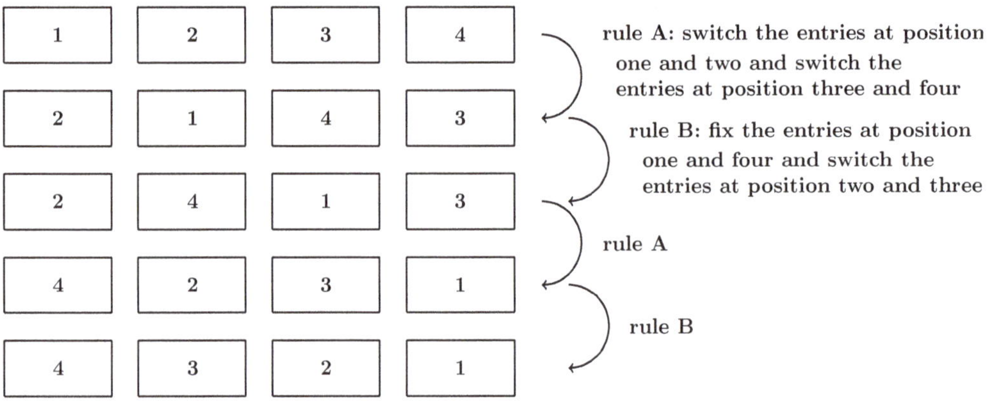
1
2
3
4
2
1
4
3
2
4
1
3
4
2
3
1
4
3
2
1
rule A: switch the entries at position one and two and switch the entries at position three and four
rule B: fix the entries at position one and four and switch the entries at position two and three
rule A
rule B

Fig. 2.3. Scheme of permutations

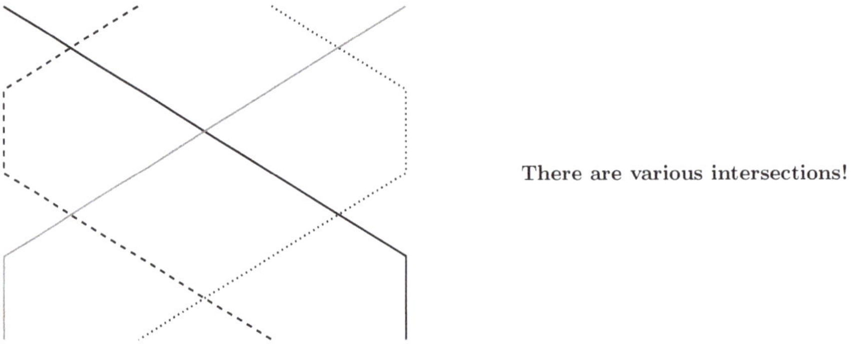
There are various intersections!

Fig. 2.4. The trajectories of the above scheme

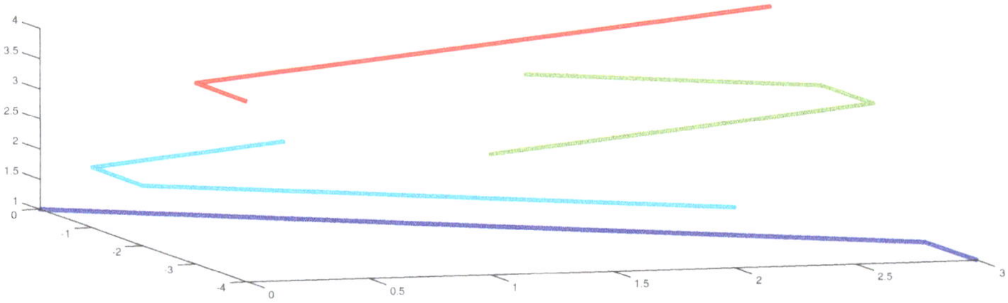

Fig. 2.5. 3-dimensional “resolution”: the trajectories no longer intersect

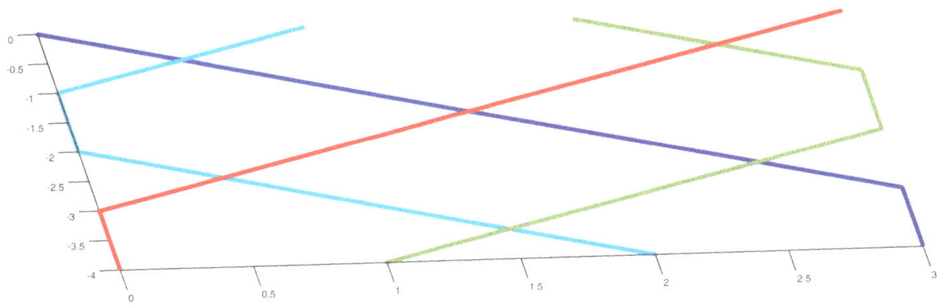

Fig. 2.6. The projection of the 3-dimensional version still looks the same

2.3 A first notion of generalized flows

After the illustrative excursion in the preceding section we come back to the ideas and properties laid out in the paragraph about the Lagrangian description of fluid motion. In particular, we will introduce a first generalized notion of an incompressible flow starting from the observation how incompressibility can alternatively be formulated in combination with the "splitting" into a horizontal part and an additional "microstructure", see property (2.1).

Definition 2.4. *Let D be the closure of a bounded, regular, open subset of $\mathbb{R}^n$ (or alternately a periodic box), let $Z = [0, 1]$ and fix a time interval $[0, T]$.*
A ***generalized incompressible flow*** *is a Borel map*

$$[0, T] \times D \times Z \ni (t, a, z) \mapsto \xi(t, a, z) \in D,$$

which is incompressible in the sense that it holds

$$\int_Z \int_D f(\xi(t, a, z))\, da dz = \int_D f(a)\, da,$$

for all continuous functions $f \in C(D)$.

Remarks 2.5.

i) This is a generalization of the hydrostatic situation we discussed previously, where D and Z were respectively the "horizontal part" and the "vertical part" of the original 3-dimensional domain.
ii) This notion is similar to the notion of Young measures in homogenization theory.

Apart from giving a generalized notion of incompressible flow, one can also give a generalized notion of a solution to the Euler equations.

Definition 2.6. *A generalized incompressible flow is a* ***generalized solution to the Euler equations*** *if there exists a sufficiently smooth $p : [0, T] \times D \to \mathbb{R}$ such that for all $t \in (0, T)$, for all $a \in D$ and for all $z \in Z$ it holds in an appropriate sense*

$$\partial_{tt}\xi(t, a, z) = -(\nabla p)(t, \xi(t, a, z)). \tag{gE}$$

Concerning the smoothness required for the pressure field to give a sense to the later ODE, let us briefly review the most important recent results on ODEs - of both first and second order - with low regularity vector fields.

i) *DiPerna-Lions*
   In their fundamental paper [22] DiPerna and Lions study the ODE

$$\frac{d\xi}{dt} = v(\xi) \tag{DpL}$$

   under the additional assumption that

$$v \in W^{1,1}_{loc}(\mathbb{R}^n), \quad \nabla \cdot v = 0 \; a.e. \text{ on } \mathbb{R}^n,$$

   and

$$v \in L^p + (1 + |x|)L^\infty.$$

   The notion of solution is the one of *renormalized solutions*. This means that the ODE (DpL) will be satisfied in the following sense:
   For all $\beta \in C^1(\mathbb{R}^n, \mathbb{R}^n)$ with the property that β as well as $|D\beta(y)|(1 + |y|)$ are both bounded on $\mathbb{R}^n$, $\beta(\xi) \in L^\infty(\mathbb{R}; L^1_{loc})$ and it holds

$$\frac{d}{dt}\beta(\xi) = D\beta(\xi) \cdot v(\xi) \text{ in } \mathcal{D}'(\mathbb{R} \times \mathbb{R}^n),$$

   and

$$\beta(\xi)|_{t=0} = \beta(y) \text{ on } \mathbb{R}^n.$$

   Their results about existence, uniqueness and stability are derived form analogous results for (linear) transport equations $u_t - v \cdot \nabla u = 0$.
   In addition, they study as well the case when the vector field v may depend on time and may be not divergence-free.

ii) *Bouchut*
   In Bouchut's article [4] the object of study is the Vlasov equation

$$\partial_t f + b \cdot \nabla_\xi f + \nabla_b \cdot (E(t, \xi)f) = 0,$$

   where f is the density of particles, E is the force field and the particles satisfy the mechanical laws

$$\frac{d\xi}{dt} = b \quad \text{and} \quad \frac{db}{dt} = E(t, \xi).$$

The crucial assumption in the cited paper is the assumption that E belongs to BV, the space of functions of bounded variation.
So, again by means of characteristics the study of the Vlasov equation can be interpreted as studying

$$\frac{d^2\xi}{dt^2} = E(t, \xi),$$

with a right hand side belonging to BV.

iii) *Ambrosio*
In Ambrosio's article [1] the theory of DiPerna-Lions is extended to the case where the right hand side of (DpL) in now a function of bounded variation, BV.

iv) *Crippa-De Lellis*
In their seminal collaboration [21] Crippa and De Lellis provide an alternative approach to the theory of DiPerna-Lions by careful a-priori-estimates for the Lagrangian formulation.

They work with the notion of *regular Lagrangian flow*. Such a flow is defined as follows:
Let v be a vector field in $L^1_{loc}([0,T] \times \mathbb{R}^n; \mathbb{R}^n)$. Then a map $\xi : [0,T] \times \mathbb{R}^n \to \mathbb{R}^n$ is called a regular Lagrangian flow for the vector field v if the two conditions below are satisfied
i) For a.e. $a \in \mathbb{R}^n$ the map $t \mapsto \xi(t, a)$ is an absolutely continuous integral solution of

$$\frac{d\xi}{dt} = v(t, \xi(t)) \quad \text{for } t \in [0,T] \quad \text{with } \xi(0) = a.$$

ii) There exists a constant L - called the compressibility constant of the flow ξ - independent of t such that for the Lebesgue measure we have

$$|\xi(t, \cdot)^{-1}(\mathcal{B})| \le L|\mathcal{B}| \quad \text{for every Borel set } \mathcal{B} \subset \mathbb{R}^n.$$

v) *Champagnat-Jabin*
In the article [17] the authors Champagnat and Jabin look at the second order situation

$$\frac{d^2\xi}{dt^2} = v(t, \xi) \quad \text{where } v \in H^{3/4} \cap L^\infty.$$

By a combination of the approach of Crippa-De Lellis and an averaging lemma they prove existence and uniqueness of solutions to the initial value problem.

2.4 Generalized flows as probability measures on path

A slightly different approach comes from a probabilistic point of view, namely from the idea to see generalized flows as probability measures on paths. This approach was used by the first author in [7] (see also [28], [10], [29] and [9]).
This point of view is reminiscent of the notion of path integral in quantum mechanics where the action is “averaged over all possible path” (see e.g. [3]).

Although the concept of probability measures on paths is well spread in probability, e.g. in Wiener processes related to Brownian motion, the strategy which we will follow here will be quite different.

2.4.1 Some preliminaries

Before we come to the heart of matter, let us on one hand recollect some useful background information and on the other hand fix the notation used in what follows.

In the sequel, $D \subset \mathbb{R}^n$ will denote the closure of a bounded open set, which often is in addition assumed to be convex.
It is also possible to work in the periodic setting, $D = \mathbb{R}^n/\mathbb{Z}^n$. But in this latter case some of the definitions below have to be modified.

Moreover, we fix a time $T > 0$ and use a different viewpoint on the classical Euler equations introduced in the first section. The problem there was an initial value problem where were given both the initial positions and the initial velocities of the fluid parcels. Here, we will rather fix the positions of fluid parcels at both the initial time $t = 0$ and the final time T.
Notice that, in geophysical fluid mechanics, through the “data assimilation” technique, the specialists of weather forecasting always take into account all available observations made at intermediate times, and, therefore, definitely do not solve an initial value problem!

We introduce the set of all paths from $[0, T]$ to D,

$$\Omega = D^{[0,T]} = \left\{[0, T] \ni t \mapsto \xi(t) \in D\right\},$$

which is compact by Tychonoff’s theorem.

Then:

- By $C(\Omega)$ we denote all continuous functions on Ω.

- Following the Riesz method, we denote $C(\Omega)^*$ the space of continuous linear forms on $C(\Omega)$ (duality).
- We introduce $C_{\text{fin}}(\Omega)$, the set of all continuous functions of finite type on Ω. More precisely, $F \in C_{\text{fin}}(\Omega)$ if there exist

$$\begin{cases} \text{i) } N \in \mathbb{N}/\{0\}, 0 \le t_1 < t_2 < \cdots < t_N \le T, \\ \text{ii) } f \in C(D^N), \end{cases}$$

such that

$$F(\xi) = f(\xi(t_1), \xi(t_2), \ldots, \xi(t_N)).$$

It holds that C_{fin} is a dense subset of $C(\Omega)$ (due to the Stone-Weierstrass-theorem).
- Furthermore, by $\text{Prob}(\Omega)$ we denote the probability measures on Ω. More precisely, $\mu \in \text{Prob}(\Omega)$ if μ belongs to $C(\Omega)^*$ and in addition it holds

$$\begin{cases} \text{i) } \langle \mu, F \rangle \ge 0, \ \forall F \ge 0, \ F \in C(\Omega), \\ \text{ii) } \langle \mu, 1 \rangle = 1. \end{cases}$$

Note that $\text{Prob}(\Omega)$ is a compact set for the weak* topology, but it is not sequentially compact (see Proposition 2.3 in [7] and also Theorem 3.2. and the related remark at the end of section 3 of the same article).
- For a given partition of $[0, T]$, $0 \le t_1 < t_2 < \cdots < t_N \le T$ and given $\mu \in \text{Prob}(\Omega)$ we define the corresponding marginal (projection) $\mu_{t_1,\ldots,t_N}$ to be the probability measure on D^N whose action on $f \in C(D^N)$ is given by

$$\langle \mu_{t_1,\ldots,t_N}, f \rangle = \langle \mu, F \rangle,$$

where

$$F(\xi) = f(\xi(t_1), \xi(t_2), \ldots, \xi(t_N)) \ \forall \xi \in \Omega.$$

2.4.2 Alternative concept of a generalized flow

With the aid of the definitions and notations introduced in the previous subsection, we can now state the definition of a generalized incompressible flow in the sense of "probability measures on paths".

Definition 2.7. *A* ***generalized flow*** *is a probability measure in Prob(Ω).*

Now we want to give a precise meaning to the concept of kinetic energy in this new framework. Indeed, let us recall that, beyond its formulation in terms of a second order ODE, (L2), the Euler equations have the following *variational formulation* (in a smooth context):
Assume that ξ and p are smooth solutions of (L2). Then this means that for each short

enough fixed time interval $[t_0, t_1]$ the trajectory ξ minimizes the action

$$\int_{t_0}^{t_1} \int_D \frac{1}{2} |\partial_t \xi(t, a)|^2 \, da dt$$

among all volume and orientation preserving diffeomorphisms that coincide with ξ at the end points t_0 and t_1.

So, our goal is to formulate this principle of least action in our present setting.

We first introduce the concept of energy for an individual path (here we suppose D to be convex in order to keep the presentation as simple as possible):

Definition 2.8. *Let D be convex, let ξ belong to Ω and let $0 \leq t_1 < t_2 < \ldots t_N \leq T$ be a discretization of the time interval $[0, T]$ for $N \in \mathbb{N}$. Then we define the following **"energy density"***

$$e(\xi) := \sup_{0 \leq t_1 < t_2 < \cdots < t_N \leq T} \frac{1}{2} \sum_{\alpha=2}^{N} \frac{|\xi(t_\alpha) - \xi(t_{\alpha-1})|^2}{t_\alpha - t_{\alpha-1}}.$$

The function $e : \Omega \to [0, \infty]$ is lower semi-continuous (cf. Proposition 3.3 in Brenier's article [7]) and has the property that whenever the considered path ξ is discontinuous for its energy density we have $e(\xi) = \infty$ (quite obvious).

Then, the "action" of the fluid, which is just the kinetic energy integrated in time over $[0, T]$, $\int_t \int_D \frac{1}{2} |\partial_t \xi(t, a)|^2 da dt$, in the classical framework, is defined, in our new framework, as follows:

Definition 2.9. *Let D, respectively Ω, ξ and $e(\xi)$ be as in the preceding definition and assume that $\mu \in Prob(\Omega)$. Then the corresponding **"energy action"** is given by*

$$A(\mu) = \int_\Omega e(\xi) \, d\mu(\xi) \in [0, \infty].$$

As in the case of the "energy density", the "energy action" as well is lower semi-continuous on Prob(Ω) (again cf. Proposition 3.3 in [7]).
Note that this fact guarantees the existence of an action minimizing path whenever there is at least one path of finite action.

Two other important features of the newly introduced "energy density" and "energy action" are the following ones:

Proposition 2.10. *Let R be an arbitrary positive real number.*
Then for this fixed constant R the set

$$Prob_R(\Omega) = \{\mu \in Prob(\Omega) \text{ such that } A(\mu) \leq R\}$$

is weak- sequentially compact.*

A proof of this fact can be found in the Appendix of [9].

Lemma 2.11. *Let ξ be a member of Ω.*
Then the "energy density" of ξ, $e(\xi)$ is finite if and only if

$$\begin{cases} i) \ \xi \in C^{1/2}([0,T], D), \\ ii) \ \xi \in H^1([0,T], \mathbb{R}^n). \end{cases}$$

If the two above conditions are satisfied, the energy density $e(\xi)$ is given by

$$e(\xi) = \frac{1}{2} \int_0^T |\xi'(t)|^2 \, dt.$$

A proof of this fact is included in part (ii) of the proof of Proposition 3.3 in [7].

The above lemma implies already a minimal regularity of a generalized flow with finite "action".

Up to now, we have introduced a new notion of generalized flows. But, as we have seen in the second section, the concept of flow alone is not yet the whole story. It is equally crucial to take into account the concept of incompressibility. This in fact is our next goal.

Definition 2.12. *A measure $\mu \in Prob(\Omega)$ with the property that for all $t \in [0,T]$ it holds*

$$\mu_t = \mathcal{L}|D,$$

where $\mathcal{L}$ is the renormalized Lebesgue measure (i.e., the Lebesgue measure divided by the volume of D), is said to be ***incompressible****.*

An interesting relation to classical incompressible flows is the following theorem which states roughly speaking that generalized incompressible flows can be seen as limits of classical incompressible flows. A proof of this statement can be found in Shnirelman's article [28]. Other approaches to the same finding are given by Brenier in [11] and by Ambrosio-Figalli in [2].

Theorem 2.13. *Assume that $n \geq 2$, that D is a nice convex domain (or the flat torus) and let μ be a generalized incompressible flow with finite action $A(\mu) < \infty$.*

Then there exists a sequence of classical incompressible flows

$$g_n : [0, T] \times D \ni (t, x) \mapsto g_n(t, x) \in D,$$

such that the corresponding generalized flows μ_n *defined by*

$$\langle \mu_n, F \rangle = \int_D F[t \to g_n(t, x)]\, dx$$

converges, i.e.,

$$\langle \mu_n F \rangle \to \langle \mu, F \rangle,$$

and also the corresponding actions converge, i.e.,

$$A(\mu_n) \to A(\mu).$$

Last but not least, in our context the time-boundary conditions are included by fixing the marginal $\mu_{0,T}$.

Then we can finally state the following "action minimization problem" (also called "minimal geodesic problem" or "shortest path problem"):
For any given time-boundary condition $\eta \in \mathrm{Prob}(D \times D)$ we want to minimize $A(\mu)$ among all $\mu \in \mathrm{Prob}(\Omega)$ with the same time-boundary, i.e., these $\mu \in \mathrm{Prob}(\Omega)$ such that

$$\mu_{0,T} = \eta.$$

In order to have a reasonable minimization problem, we should have an appropriate corresponding compact set. In our case, in fact, we have that the set

$$\mathrm{Prob}_{I,\eta}(\Omega) = \{\mu \in \mathrm{Prob}(\Omega)\ |\mu \text{ is incompressible and } \mu_{0,T} = \eta\}$$

is a compact set (again see [7]).
Note that in order to get a non-empty set from this construction, it necessarily has to hold

$$\int_D \int_D f(x)\eta(dx, dy) = \int_D f(x)\, dx = \int_D \int_D f(y)\eta(dx, dy).$$

A η satisfying this condition is called **doubly stochastic measure** or **polymorphism**. A more detailed discussion can be found e.g. in the first two sections of [7].

And what about the solvability of this "minimal geodesic problem"?
Actually, the answer is positive.

Theorem 2.14. *For every doubly stochastic* η *there exists always a solution* μ_{opt} *to the problem of minimizing the action* A.

This result can be found in [7].

After this presentation of the set-up for all our further considerations, let us go back to our former concept of generalized flows.
In the first section we defined generalized flows as maps

$$\Xi : [0, T] \times D \times Z \ni (t, a, z) \mapsto \Xi(t, a, z) \in D,$$

with the additional incompressibility property

$$\int_Z \int_D f(\Xi(t, a, z))\, dadz = \int_D f(x)\, dx \quad \forall f \in C(D).$$

Starting from that, we can obtain by duality a generalized flow according to the second definition from this section in the following way:
For all $F \in C_{fin}(\Omega)$ we set

$$\langle \mu, F \rangle = \int_Z \int_D f(\Xi(t_1, a, z), \dots, \Xi(t_n, a, z))\, dadz.$$

And moreover, the action is given by

$$A(\mu) = \frac{1}{2} \int_0^T \int_Z \int_D |\partial_t \Xi(t, a, z)|^2\, dtdadz.$$

Next, we look at a fixed time t_0. Then, according to what we have seen so far, we have

$$\begin{aligned} \langle \mu_{t_0}, f \rangle &= \langle \mu, F \rangle \quad \text{where } F(\xi) = f(\xi_{t_0}) \\ &= \int_Z \int_D f(\Xi(t_0, a, z))\, dadz = \int_D f(x)\, dx. \end{aligned}$$

This shows that, in fact, the μ associated to Ξ is incompressible.

Last, but not least, let us consider a particular choice of our time-boundary condition η:
We require

$$\begin{cases} \text{i) } \Xi(0, a, z) = a & \text{just the initial position, the identity map,} \\ \text{ii) } \Xi(T, a, z,) = h(a) & \text{for a given } h : D \to D \\ & \text{independent on the microstructure } z \\ & \text{and compatible with the incompressibility condition.} \end{cases}$$

This latter compatibility condition with the incompressibility requirement reads a follows

$$\int_D f(h(a))\, da = \int_D f(x)\, dx \quad \text{for all } f \in C(D),$$

i.e., h is a volume preserving map in the measure theoretic sense.
With this assumption we then get

$$\begin{aligned}\langle \mu_{0,T}, F\rangle &= \int_D \int_Z f(a, h(a))\, da dz \\ &= \int_D f(a, h(a))\, da.\end{aligned}$$

As we have seen in the preceding section, once we have a notion of generalized flow at hand we can go ahead and ask for a suitable notion of generalized solution to the Euler equations starting from the generalized flow concept. In fact, a somehow natural notion arising from the above concept of generalized flow is the following definition of generalized solution

Definition 2.15. *A generalized flow μ is a* ***generalized solution to the Euler equations*** *if there exists a suitably regular pressure*

$$p : [0, T] \times D \to \mathbb{R},$$

such that for μ almost every path ξ it holds

$$\frac{d^2\xi}{dt^2} = -(\nabla p)(t, \xi(t)).$$

As one of the fundamental results, the following uniqueness result for generalized solutions to the Euler equations hold, see for instance Theorem 5.1 in [7].

Theorem 2.16. *Let μ be a generalized solution to the Euler equations such that the corresponding pressure p of regularity C^2 in x with the following uniform bound on its Hessian*

$$D_x^2 p(t, x) \leq K \quad \forall t \in [0, T]\ \forall x \in D$$

for some $K \in \mathbb{R}$, i.e.,

$$\zeta D_x^2 p(t, x)\zeta \leq K|\zeta|^2 \quad \forall \zeta \in \mathbb{R}^n,$$

then μ is the unique action minimizer on any time interval $[t_0, t_1] \subset [0, T]$ provided that

$$|t_1 - t_0|^2 K \leq \pi^2,$$

and μ_{t_0,t_1} is fixed.

Note that this result can be read in the sense that the generalized solution is “locally the shortest path”.

Remarks 2.17.

i) *Answers to the question how this concept of generalized solutions to the Euler equations is related to classical or weak solutions of the Euler equations are given in Shnirelman's article [29] (theorem 3.2) an in [7].*
ii) *The ideas presented above do not restrict to problems from fluid dynamics. Also problems from e.g. electrodynamics lead to PDEs which can be studied with a similar technique (see e.g. [12]).*

2.5 Interpretation and examples

In this section we will present some examples of generalized flows according to the concept introduced in the preceding section.
In the following examples we will see the impact of "dimensional collapse" with "mixing in the vertical direction".
More precisely, the situation under consideration is the following:
The domain D is the product of a horizontal domain $D_H \subset \mathbb{R}^2$ (e.g. the unit square) and the vertical interval $H = [-1, 1]$, i.e., $D = D_H \times H$, and we want to study possible "mixings" of a particular kind.
Our goal is to find a generalized flow which acts only in the "vertical direction", i.e., such that its projection to D_H is the identity map.
This situation can also be interpreted as a "fluid layer".

As explained in [14], in the classical theory it is not possible to have fluid motion only in the vertical direction. But as we will see, in the relaxed setting of generalized flows this actually can be achieved.

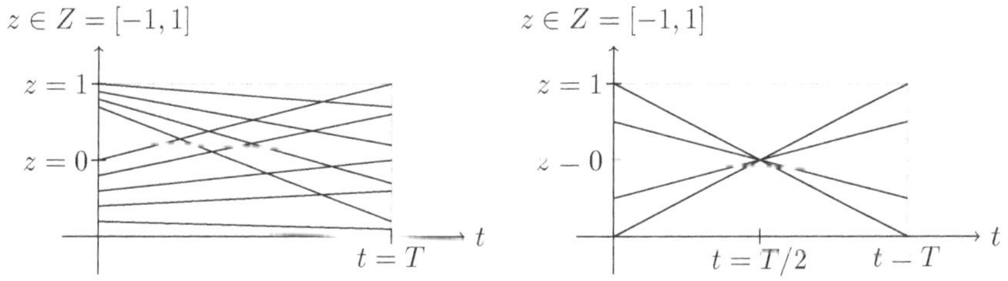

Fig. 2.7. Illustration of the possible trajectories in the vertical direction for a fixed point in D_H

Before we go into some details of concrete examples, let us illustrate the phenomena which should be captured. To this end, the pictures above show how possible

trajectories should look like.
In the figures below we fix a point in D_H and just depict the time-evolution in the third, vertical direction. Note that these pictures are illustrations and not precisely calculated trajectories. (Numerically) calculated trajectories are provided e.g. in [14].
Next, we will become more precise by providing some concrete examples of generalized flows with just one space variable.

Example 1
In this first example, we set $T = 1$ and we want to go from the identity map at time 0 to the following (modified) "doubling map" at time $T = 1$:

$$\eta(x_3) = \begin{cases} 2\sqrt{x_3} - 1, & 0 \le x_3 \le 1, \\ 1 - \sqrt{-4x_3}, & -1 \le x_3 < 0, \end{cases}$$

where x_3 denotes the vertical coordinate $x_3 \in Z$, see Figure 2.8.

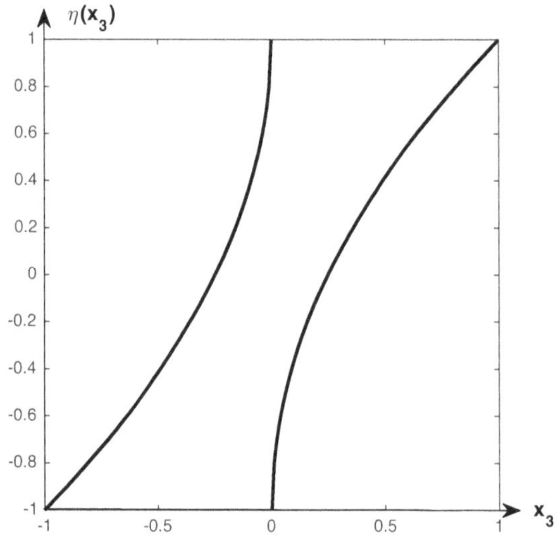

Fig. 2.8. Final configuration η

As we can observe just from the prescribed final configuration, half of the domain covers the whole target (see also Figure 2.10 below).

Then, it turns out that

$$\Xi(t, x_1, x_2, x_3) = (x_1, x_2, \bar{\eta}(t, x_3)),$$

where

$$\bar{\eta}(t,x_3)=\begin{cases} t^{2/3}\eta(x_3\cdot t^{-2/3}), & |x_3|\le t^{2/3},\\ x_3, & \text{otherwise,}\end{cases}$$

is a generalized flow according to Definition 2.4.
These trajectories are shown in the Figure 2.9.
Actually, if we look at a more refined picture, one can see that half of the domain of x_3, i.e., [0, 1] or [−1, 0] cover the whole interval [−1, 1] in the final configuration. More-

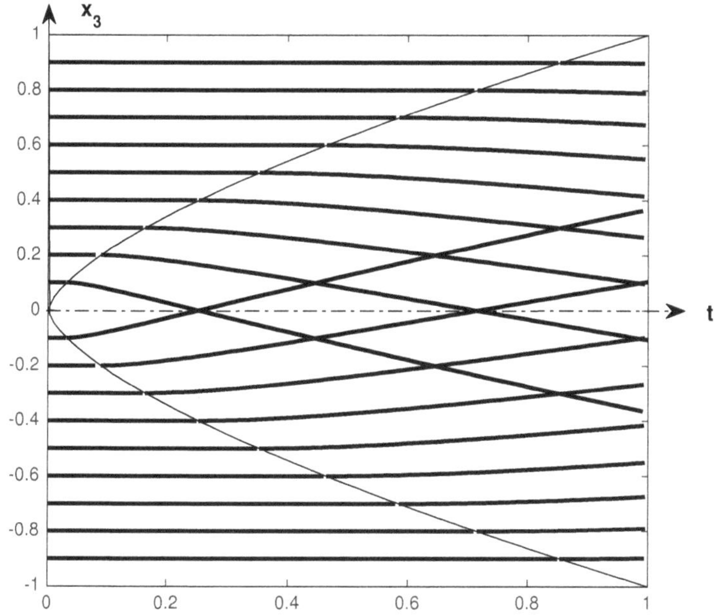

Fig. 2.9. Trajectories Ξ (bold) and the curve $x_3 = t^{2/3}$

over, this flow is also a generalized solution to the Euler equations in the sense of Definition 2.6.
In fact, it holds that (gE) is true, i.e.,

$$\partial_{tt}\Xi(t,x_1,x_2,x_3)=(0,0,(\partial_{x_3}\bar{p})(t,\Xi(t,x_1,x_2,x_3))),$$

where

$$\bar{p}(t,x_3)=-\frac{1}{9}t^{-2}(t^{4/3}-x_3^2)_+.$$

What about the formulation in terms of probability measures, i.e., in terms of Definitions 2.7 and 2.15?
If for F of finite type we set

$$\langle\mu,F\rangle=\int f(\Xi(t_1,x),\dots,\Xi(t_N,x))\,dx,$$

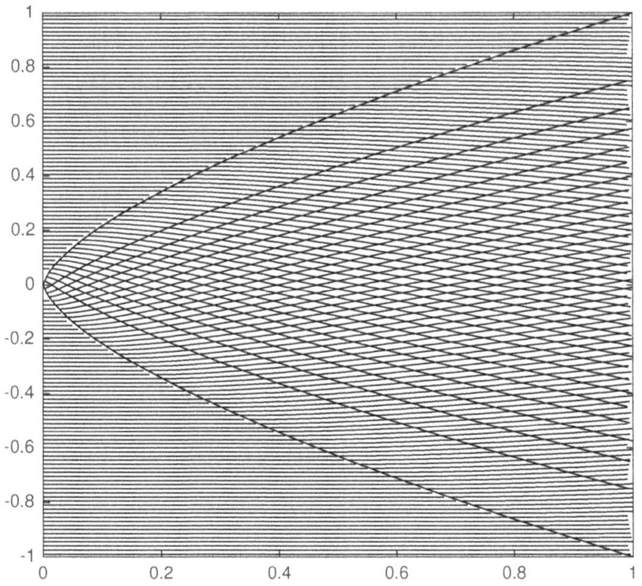

Fig. 2.10. More trajectories

one can show that this μ is a generalized solution of the Euler equations and is action minimizing.

Further details and results in an broader context can be found in [15] (see in particular Theorem 4.1).
As a matter of fact, this example (with a different but equivalent formulation) goes back to the article of Duchon-Robert [23].

Example 2
In this second example, the final datum is

$$\eta(x_3) = -x_3$$

and the domain D is the same as in the first example, but now the final time is $T = \pi$. In addition, just from the beginning we impose the pressure, more precise we look at

$$p(t, x_1, x_2, x_3) = \frac{1}{2}{x_3}^2.$$

From this, we can immediately deduce that in order to have a generalized solution to the Euler equations, it is necessarily that

$$\partial_{tt}\Xi = (0, 0, -\Xi_3),$$

and so in particular

$$\Xi_3(t) = A\cos(t) + B\sin(t),$$

where A and B depend on the particular trajectory we are looking at (and should satisfy the incompressibility condition).
This problem actually can be solved in the following way:
If we introduce an additional variable, an additional "microstructure" $z \in [0, 1]$ then

$$\Xi(t, x_1, x_2, x_3, z) = (x_1, x_2, x_3 \cot(t) + \sqrt{1 - x_3^2} \cos(2\pi z) \sin(t))$$

is the desired solution.
Observe that the acceleration - the pressure - does not depend on the additionally introduced "microstructure" z.
Nevertheless, the situation here is quite different from the situation we have encountered in the first example. Here, there is no longer just one trajectory connecting the initial configuration - the identity - with the final configuration given by the $\eta(x_3) = -x_3$ but there is a whole family of trajectories connecting the two given configurations (foliation of a dynamical system). In the picture below, this is shown for fixed $x_2 = 0.25$. Each curve corresponds to a different value of z.

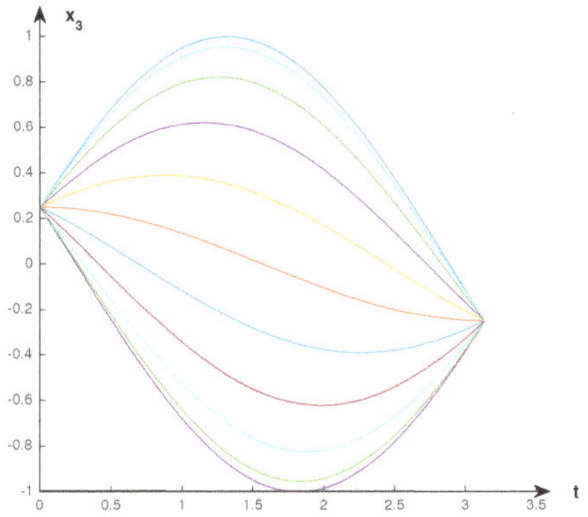

Fig. 2.11. Splitting of the particle trajectory

Moreover, this solution minimizes the action (see for instance Proposition 6.3 in [7]) and is the unique action minimizer according to Theorem 2.16.

2.6 The dual problem

So far, we have seen the action minimizing problem

$$\inf_{\mu \in AD} A(u),$$

where AD was the following set of admissible competitors

$$AD = \left\{\mu \in \text{Prob}(\Omega) \mid \mu_t \text{ is the Lebesgue measure } \forall t \in I \text{ and } \mu_{0,T} = \eta\right\},$$

where

$$\eta \text{ is doubly stochastic.}$$

In the spirit of Kantorovich's approach to the Monge optimal transport problem, our minimization problem too can be reformulated as a maximization problem through a duality technique.

Indeed, the following theorem (see for instance [9]) holds.

Theorem 2.18. *It holds*

$$\inf_{\mu \in AD} A(\mu) = \sup_{f \in ad} \langle \mu^{ref}, f \rangle,$$

where μ^{ref} is an arbitrary element of AD of finite action and

$$ad = \left\{ f \in C(\Omega) | \langle \mu - \mu^{ref}, f \rangle = 0 \, \forall \mu \in AD \text{ and } f(\xi) \leq e(\xi) \, \forall \xi \in \Omega \right\}.$$

This theorem is crucially based on the Legendre-Fenchel-Moreau transforms, the Fenchel-Rockafellar duality theorem and a time discretization argument. For a presentation of the first two ingredients, the interested reader is referred to Brezis' monograph [16].

The advantage of this reformulation of our problem is that thanks to this alternative formulation it is possible to prove that for given time-boundary data η there exists an optimal flow, μ_{opt} and that in addition for this optimal flow, there is a corresponding "pressure", p_{opt} (roughly speaking this flow is a generalized solution to the Euler equations) which is basically determined only by the time-boundary datum η. These facts are proved in Theorems 1.1 and 1.2 in [9].
What concerns the regularity of this optimal pressure, it holds that the pressure gradient is a locally bounded measure on $(I \times D)^\circ$ (see [11], Theorem 1.2). Additionally, further integrability of the optimal pressure was proved by Ambrosio-Figalli in [2] (Theorem 6.3).

2.7 Relation to optimal transport and Monge-Ampère

So far, the time variable t was always considered continuous (the marginals $\mu_{t_1,\ldots,t_N}$ took into account only finitely many points in time, but the time variable itself was continuous).

Now, we want to look at discrete version of what we have seen above.
In order to do so, we pick $N \in \mathbb{N}$ and denote the corresponding time-steps by $t_0 = 0 < t_1 = 1/N < \cdots < t_N = 1$. Thus we have a time-discretization of the unit interval in time. Note that in this way we have $N + 1$ points in time.
Again, for $\mu \in \mathrm{Prob}(\Omega)$ we prescribe the boundary condition $\mu_{0,1} = \eta$ and require that

$$\mu_t = \mathcal{L} \quad \text{for all } t = t_k,\ k = 0, \ldots, N,$$

where $\mathcal{L}$ denotes the Lebesgue measure.
And the action which we want to minimize is now the discrete analogue

$$A_{discrete}(\mu) = \frac{1}{2} \int_{D^{N+1}} \sum_{k=1}^{N} \frac{|\xi(t_k) - \xi(t_{k-1})|^2}{(t_k - t_{k-1})}\, d\mu(\xi) = \int_{D^{N+1}} F_N(\xi)\, d\mu(\xi).$$

This time-discrete problem belongs to the class of "multi-marginal optimal transport problems".

This is a very active field of research, with many applications, for instance to quantum chemistry (classical approximation to the density functional theory) and information theory (computation of the barycenter of a given collection of measures) for "big data" analysis.

Let us consider the simplest case $N = 2$.
The action functional we want to study further reads - up to a constant factor -

$$\int |\xi(t_1) - \xi(t_0)|^2 + |\xi(t_2) - \xi(t_1)|^2\, d\mu(\xi).$$

In addition we simplify our point of view in so far that we are given a *volume-preserving map* instead of a measure:

$$\eta(dx, dy) = \delta(y - h(x))\, dx \quad \text{where } h \text{ is a volume-preserving map.}$$

In this slightly modified setting, our minimization problem is finally rephrased as follows:

$$\min \int_{D^2} \left(|a - b|^2 + |b - h(a)|^2\right) d\nu(a, b), \tag{2.2}$$

with

$$\nu \quad \text{doubly stochastic,}$$

since

$$\int F_2(x_0, x_1, x_2)\, d\mu_{0,1,2}(x_0, x_1, x_2) = \int_{D^2} F_2(a, b, h(a))\, d\nu(a, b).$$

The argument which has to be minimized in (2.2) can be further rewritten, which leads us to the following equivalent minimization problem

$$\min \int_{D^2} \left(\left| b - \frac{a + h(a)}{2} \right|^2 \right) d\nu(a, b) = \min \int_{D^2} \left(\left| b - H(a) \right|^2 \right) d\nu(a, b)$$

$$= \min \int_{D^2} f(b, H(a))\, d\nu(a, b) \quad \text{where } f(b, c) = |b - c|^2,\ H(a) = (a + h(a))/2.$$

Adapting further this strategy one finally finds the minimization problem (note that the marginals have to be adapted)

$$\min \int_{D^2} |b - c|^2\, d\lambda(a, b), \tag{2.3}$$

where

$$\int_{D^2} f(b, H(a))\, d\nu(a, b) = \int_{D^2} f(b, c)\, d\lambda(a, c) \quad \text{for all } f \in C(D^2).$$

In this particular situation, we have

$$\begin{aligned} \int f(c)\, d\lambda(b, c) &= \int f(H(a))\, d\nu(a, b) = \int f(H(a))\, da \\ &= \int f(y)\, \rho(dy) \end{aligned}$$

and

$$\begin{aligned} \int f(b)\, d\lambda(b, c) &= \int f(b)\, d\nu(a, b) \\ &= \int f(b)\, db. \end{aligned}$$

But then, our minimization problem (2.3) is nothing else than the *optimal transport problem* of moving the normalized Lebesgue measure λ on D to the given measure $\rho \in \mathrm{Prob}(D)$ with quadratic cost function $|a - b|^2$.
One of the best known results then is the following theorem (see e.g. [6])

Theorem 2.19. *If ρ is absolutely continuous with respect to the Lebesgue measure then the optimal λ in (2.3) is unique and of the form*

$$\int f(b, c,)\, d\lambda(b, c) = \int f(\nabla\varphi(c), c)\, \rho(dc) = \int f(\nabla\varphi(c), c)\, \tilde{\rho}(c)\, dc,$$

where φ is a convex function.

From this theorem we can formally deduce that

$$\begin{aligned}\int f(b)\, d\lambda(b,c) &= \int f(b)\, db \\ &= \int f(\nabla\varphi(c))\, \tilde{\rho}\, dc \\ &= \int f(\nabla\varphi(c))\, det\, D^2\varphi(c)\, dc,\end{aligned}$$

which implies that

$$det\, D^2\varphi(c) = \tilde{\rho}(c).$$

But this latter equation is exactly the *Monge-Ampère equation*.

Further results and a more detailed discussion of the relations between the concept of action minimizer we have see above and the optimal theory on one hand and the Monge-Ampère equation on the other hand can be found in Brenier's articles [6] and [8].

Acknowledgment: The authors would like to thank the organizers, Anna Mazzucato and Gianluca Crippa, for the very interesting Summer School in Levico Terme and the Fondazione Bruno Kessler und the CIRM for their administrative support and hospitality.
The second author would like to thank the organizers in addition for having given her the opportunity to report the present lecture notes.

Bibliography

[1] L. Ambrosio. Transport equation and Cauchy problem for BV vector fields. *Invent. Math.*, 158(2):227–260, 2004.
[2] L. Ambrosio and A. Figalli. Geodesics in the space of measure-preserving maps and plans. *Arch. Ration. Mech. Anal.*, 194(2):421–462, 2009.
[3] G. Baym. *Lectures on quantum mechanics*. Addison-Wesley, 1990.
[4] F. Bouchut. Renormalized solutions to the Vlasov equation with coefficients of bounded variation. *Arch. Ration. Mech. Anal.*, 157:75–90, 2001.
[5] Y. Brenier. Un exemple d'approche algorithmique en recherche mathématique. www.cmls.polytechnique.fr/perso/brenier/pagesweb-cmls/Brenier-FTFNTPMP-2011.pdf.
[6] Y. Brenier. Décomposition polaire et réarrangement monotone des champs de vecteurs. *C. R. Acad. Sci. Paris*, 305:805–808, 1987.
[7] Y. Brenier. The least action principle and the related concept of generalized flows for incompressible inviscid fluids. *J. of the A.M.S*, 2:225–255, 1989.
[8] Y. Brenier. Polar factorization and monotone rearrangement of vector value functions. *Comm. Pure and Applied Maths*, 64:375–417, 1991.
[9] Y. Brenier. The dual least action problem for an incompressible flow. *Arch. Ration. Mech. Anal.*, 122:323–351, 1993.
[10] Y. Brenier. A homogenized model for vortex sheets. *Arch. Ration. Mech. Anal.*, 138(4):319–363, 1997.

[11] Y. Brenier. Minimal geodesics on groups of volume-preserving maps and generalized solutions of the Euler equations. *Comm. Pure and Applied Maths*, 52:411–452, 1999.
[12] Y. Brenier. Some geometric PDEs related to hydrodynamics and electrodynamics. *Proceedings of the International Congress of Mathematicians*, 3, 2002.
[13] Y. Brenier. Remarks on the derivation of the hydrostatic Euler equations. *Bull. Sci. Math.*, 127(7):585–595, 2003.
[14] Y. Brenier. Generalized solutions and hydrostatic approximation of the Euler equations. *Physica D: Nonlinear Phenomena*, 237(14-17):1982–1988, 2008.
[15] Y. Brenier. Remarks on the minimizing geodesic problem in inviscid incompressible fluid machanics. *Calc. Variations and PDEs*, 47(1):55–64, 2013.
[16] H. Brezis. *Analyse fonctionnelle*. Masson, 1983.
[17] N. Champagnat and P.-E. Jabin. Well-posedness in any dimension for Hamiltonian flows with non BV force terms. *Communications in Partial Differential Equations*, 35(5):786–816, 2010.
[18] P. Constantin. An Eulerian-Lagrangian approach for incompressible fluis: Local theory. *Journal of the American Mathematical Society*, 14:263–278, 2001.
[19] P. Constantin. An Eulerian-Lagrangian approach to the Navier-Stokes equations. *Comm. Math. Phys.*, 216, 2001.
[20] G. Crippa. *The flow associated to weakly differentiable vector fields*. Theses of Scuola Normale Superiore di Pisa, 2009.
[21] G. Crippa and C. DeLellis. Estimates and regularity results for the DiPerna-Lions flow. *J. Reine Angew. Math.*, 616:15–46, 2008.
[22] R. J. DiPerna and P. L. Lions. Ordinary differential equations, transport theory and Sobolev spaces. *Invent. Math.*, 98:511–547, 1989.
[23] J. Duchon and R. Robert. Relaxation of the Euler equations and hydrodynamic instabilities. *Quarterly Appl. Math*, 50:235–255, 1992.
[24] L. Euler. Principes généraux du mouvement des fluides. *Mémoires de l'académie des sciences de Berlin*, 11:274–315, 1757.
[25] L. C. Evans. *Partial Differential Equations*. American Mathematical Society, 1998.
[26] L. D. Landau and E. M. Lifshitz. *Fluid Mechanics*. Pergamon Press, 1987.
[27] A. L. Majda and A. L. Bertozzi. *Vorticity and Incompressible Flow*. Cambridge University Press, 2001.
[28] A. Shnirelman. Generalized fluid flows, theirs approximation and applications. *Geom. Funct. Anal.*, 4(5):586–620, 1994.
[29] A. Shnirelman. *Weak solutions of incompressible Euler equations*. North-Holland, 2003.

Peter Constantin*, Laura Gioia Andrea Keller, and Camilla Nobili

Existence, uniqueness, regularity and long time behavior of hydrodynamic evolution equations

Abstract: In these lecture notes we present some recent results regarding existence, uniqueness and regularity for two well-known mathematical problems arising from hydrodynamics. Although the problems are very different and therefore need a specific analysis, it turns out that in both problems at least one of the tools needed comes from harmonic analysis, for instance the study of fractional Laplace operators and singular integrals.

The first part deals with a class of problems which can be described by systems of non-linear partial differential equations consisting of incompressible Navier-Stokes or Stokes equations coupled to other field equations. The most important examples in this class are the ideal magneto-hydrodynamics or complex fluids of Oldroyd-B type. For these problems we present some results concerning existence and uniqueness obtained using a combined Lagrangian-Eulerian approach and suitable commutator estimates.

In the second part we summarize some recent developments concerning the regularity of the critical surface quasi-geostrophic equation (SQG) on the two-dimensional torus. After recalling some classical a-priori estimates and a non-linear lower bound for fractional Laplacians, we prove global (in time) regularity for the critical dissipative SQG by combining the De Giorgi iteration technique with the (above mentioned) non-linear lower bound. In particular, this result shows the existence of a compact absorbing set in $H^1(\mathbb{T}^2)$ which attracts uniformly all the dynamics.

Keywords: Ideal magneto-hydrodynamics, complex fluids of Oldroyd-B type, surface quasi-geostrophic equation, commutator estimates, Calderòn-Zygmund operators, De Giorgi iteration method, non-linear maximum principle

***Corresponding Author: Peter Constantin:** Department of Mathematics, Princeton University, U.S.A., E-mail: const@math.princeton.edu

Laura Gioia Andrea Keller: Lucerne University of Applied Sciences and Arts, Switzerland, E-mail: laura.keller@math.ethz.ch

Camilla Nobili: Departement Mathematik und Informatik, Universität Basel, Switzerland, E-mail: camilla.nobili@unibas.ch

3.1 Existence and Uniqueness results for some hydrodynamic evolution equations

In this part we present some recent results concerning the existence and uniqueness of solutions of limited smoothness for certain hydrodynamic models. In these models fluids are coupled with additional fields (representing for instance microscopic insertions or magnetic fields). We give below precise formulations, the functional framework and discuss proofs.

3.1.1 Systems studied, motivation of the approach and outline of the strategy

The partial differential equations under consideration can be written in an unified manner as Navier-Stokes equations (NS), together with the incompressibility condition (IC), coupled to equations for additional fields (C):

$$\begin{cases} \partial_t u + u \cdot \nabla u - \nu \Delta u + \nabla p = A(c), & \text{on } \mathbb{R}^n \times \mathbb{R}^+, & (NS) \\ \nabla \cdot u = 0, & \text{on } \mathbb{R}^n \times \mathbb{R}^+, & \text{(IC)} \\ \partial_t c + u \cdot \nabla c = F(\nabla u, c), & \text{on } \mathbb{R}^n \times \mathbb{R}^+, & \text{(C)} \end{cases}$$

where $u = u(x, t) \in \mathbb{R}^n$ with $x \in \mathbb{R}^n$ for $n = 2, 3$.

For example, in the case of ideal magneto-hydrodynamics (MHD) we have $c = b$, the divergence-free magnetic field

$$\begin{cases} A(c) & = b \cdot \nabla b = div(b \otimes b), \\ F(\nabla u, c) & = b \cdot \nabla u. \end{cases}$$

In the case of non-linear Oldroyd-B like systems (a model for complex fluids where a solvent interacts with particles) we have for $c = \sigma$ the stress tensor

$$A(c) = div\, \sigma,$$

and F a smooth function satisfying some growth condition (see section 3.1.2 below).

Motivation of the combined Lagrangian-Eulerian approach

We start from a purely Eulerian approach and look at the involved choice of function spaces. As usual, in order to study the uniqueness issue we take the difference of two solutions.

Thus, in the MHD case we assume that we have two solutions (u_1, b_1) and (u_2, b_2) such that $u_i \in C^{1+\alpha}(\mathbb{R}^2) \cap W^{1,p}(\mathbb{R}^2)$ and $b_i \in C^{\alpha}(\mathbb{R}^2) \cap L^p(\mathbb{R}^2)$, $i = 1, 2$. Then in particular we have

$$\begin{cases} \partial_t u_1 + u_1 \cdot \nabla u_1 - \nu \Delta u_1 + \nabla p_1 & = b_1 \cdot \nabla b_1, \\ \partial_t u_2 + u_2 \cdot \nabla u_2 - \nu \Delta u_2 + \nabla p_2 & = b_2 \cdot \nabla b_2, \end{cases}$$

and

$$\begin{cases} \partial_t b_1 + u_1 \cdot \nabla b_1 & = b_1 \cdot \nabla u_1, \\ \partial_t b_2 + u_2 \cdot \nabla b_2 & = b_2 \cdot \nabla u_2. \end{cases}$$

Taking the differences leads to

$$\begin{aligned} \partial_t(u_1 - u_2) + u_1 \cdot \nabla u_1 - u_2 \cdot \nabla u_2 - \nu\Delta(u_1 - u_2) + \nabla p &= b_1 \cdot \nabla b_1 - b_2 \cdot \nabla b_2, \\ \partial_t(b_1 - b_2) + u_1 \cdot \nabla b_1 - u_2 \cdot \nabla b_2 &= b_1 \cdot \nabla u_1 - b_2 \cdot \nabla u_2. \end{aligned}$$

The last equations can be written, denoting

$$u = u_1 - u_2, \quad b = b_1 - b_2, \quad p = p_1 - p_2$$

and

$$\bar{u} = \frac{1}{2}(u_1 + u_2), \quad \bar{b} = \frac{1}{2}(b_1 + b_2),$$

as

$$\begin{aligned} \partial_t u + \bar{u} \cdot \nabla u + u \cdot \nabla \bar{u} - \nu\Delta u + \nabla p &= div(\bar{b} \otimes b + b \otimes \bar{b}) \\ \partial_t b + \bar{u} \cdot \nabla b + u \cdot \nabla \bar{b} &= (\nabla \bar{u})b + (\nabla u)\bar{b}. \end{aligned}$$

From this reformulation, we can see that if we would try to close the estimates in the above mentioned function spaces we would run into trouble, since not all the quantities appearing in the last two equations are well behaved. In particular, the terms involving $\nabla \bar{b}$ or ∇b do not belong to the right function space.

From this, we see that a purely Eulerian approach is not the best one. Indeed, a combined Eulerian-Lagrangian method will be used, where the corresponding quantities in Lagrangian variables can be controlled in a much better way (commutator estimates).
For a review of the Eulerian and the Lagrangian approach to fluid dynamics the interested reader is referred to the notes of Brenier's lecture in the same volume.

Outline of the strategy

Before we go into some further details, we give a short outline of the strategy which will lead to the desired (local) existence and uniqueness of solutions to the coupled Navier-Stokes system (NS) - (C).

- In a first step the consideration is reduced from the non-linear Navier-Stokes systems to the *linear Stokes system* (see (S) - (S-C) below).
- The second step consists in a reformulation of the problem at hand in *Lagrangian-Eulerian variables* where the fact that the solution to the Stokes system can be written using *Duhamel's formula* is used.

- Based on the reformulation using Duhamel's formula in the preceding step, we can define a *map* (see (3.2) below) which will turn out to be a *contraction* in a suitable function space, thus providing the desired solution (existence result).
- The fact that the map introduced in the last step is a contraction relies on two important pillars:
  On one hand, it relies on a refined analysis of the involved operators and most importantly on the fine analysis of the involved *commutators*. The fact that the worst terms are commutators is crucial. This part will lead to the conclusion that the map maps to the right function space.
  On the other hand it relies on the analysis of *variational derivatives*. This second step will allow us to deduce that the map in fact is a contraction.
- Once existence is established, an additional *Lipschitz bound for the time derivative of Lagrangian particle path* allows us to deduce in addition uniqueness of the solution for given initial data (uniqueness result).
  This last step again relies on the analysis of the variational derivative of the map introduced in the third step.
- Once a detailed understanding of the situation of the Stokes system is well established, the Navier-Stokes system can be treated in a *perturbative manner*. Thus for the Navier-Stokes system existence and uniqueness in a suitable setting can be proven as well.

Remark 3.1. *A similar combined Lagrangian-Eulerian approach is also used in Constantin's articles [2] and [3].*

3.1.2 Set-up and preliminaries

In this section we explain the notation used in the sequel, make precise some assumptions and recall some important facts.

Stokes instead of Navier-Stokes

In what follows, we will not directly study the system (NS) - (C). First, we study the related problem where instead of the non-linear Navier-Stokes equations we have the Stokes equations, i.e., we study first systems of the following form

$$\begin{cases} \partial_t u - \Delta u + \nabla p = div(\sigma), & \text{on } \mathbb{R}^n \times \mathbb{R}^+, \quad (S) \\ \nabla \cdot u = 0, & \text{on } \mathbb{R}^n \times \mathbb{R}^+, \quad \text{(S-IC)} \\ \partial_t \sigma + u \cdot \nabla \sigma = F(\nabla u, c), & \text{on } \mathbb{R}^n \times \mathbb{R}^+. \quad \text{(S-C)} \end{cases}$$

Assumptions on F

We assume that $F(\nabla u, c)$ is smooth and has polynomial growth at infinity, i.e.,

$$|F(g, \xi)| \le C(|g| + |\xi|)^k \quad \text{for some } k \in \mathbb{N}.$$

A typical example for such a function F, apart from the precise one we have already seen in the case of the magneto-hydrodynamics, is the following one

$$F(\nabla u, c) = (\nabla u)c + c(\nabla u)^T - c + \nabla u + (\nabla u)^T.$$

This particular structure of F in the case of Olroyd-B fluids takes into account that these models enjoy a gauge invariance. For further explanations about the structure of the right hand side of the equation for F, the interested reader is referred to the article of Constantin and Sun [6] (see in particular equations (2.44), (2.29) and (4.1), note that c of these notes corresponds to τ in [6] and set $\rho = 1$).

Solution formula for the Stokes equations

We recall that the solution of the Stokes equations (S) with initial velocity $u_0(x) = u(x, t = 0)$ is given through Duhamel's principle (see e.g. Taylor's monograph [17]) by

$$u(x, t) = \mathbb{L}(u_0)(x, t) + \mathbb{U}(\sigma)(x, t),$$

where

$$\mathbb{L}(u_0) = e^{t\Delta} u_0,$$

$$\mathbb{U}(\sigma) = \int_0^t e^{(t-s)\Delta} \mathbb{H}\, div\, \sigma(s)\, ds,$$

and

$$\mathbb{H} = \mathbb{I} + R \otimes R,$$

with R the Riesz transforms (cf. e.g. Stein's book [15]) $R = (-\Delta)^{-1/2}\nabla$, respectively

$$\mathcal{F}(R_j f)(\xi) = i\frac{\xi_j}{|\xi|}\mathcal{F}(f)(\xi), \quad \text{where } \mathcal{F} \text{ denotes the Fourier transform.}$$

Note that in addition we have

$$\nabla u = e^{t\Delta}\nabla u_0 + \mathbb{G}(\sigma) = \mathbb{L}(\nabla u_0) + \mathbb{G}(\sigma),$$

where

$$\mathbb{G}(\sigma) = \int_0^t e^{(t-s)\Delta} \nabla\, \mathbb{H}\, div\, \sigma(s)\, ds.$$

Note that $\mathbb{U}(\sigma)$ is divergence free and that $\mathbb{L}$ preserves the property of having vanishing divergence.

Lagrangian description

The Lagrangian particle path X satisfies

$$\frac{dX}{dt} = u(X) = \mathbb{L}(u_0) \circ X + \mathbb{U}(\tau \circ X^{-1}) \circ X, \quad \text{where} \quad \tau = \sigma \circ X,$$

with initial data

$$X(a, 0) = a.$$

The variables $\tau = \tau(a, t)$ and $X = X(a, t)$ are called *Lagrangian variables.*

Our next goal is to find suitable expressions for τ and X.
Observe that, if we set

$$g(a, t) = (\nabla u)(X(a, t), t) = \mathbb{L}(\nabla u_0) \circ X + \mathbb{G}(\tau \circ X^{-1}) \circ X, \tag{3.1}$$

then we have

$$\frac{d\tau}{dt} = F(g, \tau).$$

In these variables we have

$$\begin{cases} X(a, t) & = a + \int_0^t \mathcal{U}(X(s, \tau(s))\, ds, \\ \tau(a, t) & = \sigma_0(a) + \int_0^t \mathcal{T}(X(s), \tau(s))\, ds, \end{cases}$$

where

$$\begin{cases} \mathcal{U} & = \mathbb{L}(u_0) \circ X + \mathbb{U}(\tau \circ X^{-1}) \circ X, \\ \mathcal{T} & = F(g, \tau) = F((\mathbb{L}(\nabla u_0) \circ X + \mathbb{G}(\tau \circ X^{-1}) \circ X, \tau). \end{cases}$$

For later purposes we rewrite the above expressions for the Lagrangian variables as one single map

$$\mathcal{S}(X, \tau) = \begin{pmatrix} a + \int_0^t \mathcal{U}(X(s, \tau(s))\, ds \\ \sigma_0(a) + \int_0^t \mathcal{T}(X(s), \tau(s))\, ds \end{pmatrix}. \tag{3.2}$$

We prove below that the map $\mathcal{S}$ is a contraction in a suitable space and thus leads to a solution of our problem.

Starting from these expressions, in particular the map $\mathcal{S}$, we can study variational derivatives (Gateaux derivatives) in order to prove that $\mathcal{S}$ is a contraction.
The terms which we obtain calculating variational derivatives will reveal a first instance of where we gain in using a combined Lagrangian-Eulerian approach instead of a purely Eulerian one. In fact, the expression we will get are commutators which are nicely behaved.
In order to achieve this goal, we consider a differentiable one-parameter family of paths X_ε with corresponding τ_ε and initial data $u_\varepsilon(0)$ and $\sigma_\varepsilon(0)$. We assume that $X_\varepsilon \in Lip(0, T; C^{1+\alpha,p}(\mathbb{R}^n))$, $\tau_\varepsilon \in Lip(0, T; C^{\alpha,p}(\mathbb{R}^n))$, $u_0 \in C^{1+\alpha,p}(\mathbb{R}^n)$ and

$\sigma_0 \in C^{\alpha,p}(\mathbb{R}^n)$ - the precise definitions of these spaces is given in the next paragraph below - and introduce the following notation

$$X_\varepsilon' = \frac{dX_\varepsilon}{d\varepsilon} \quad \text{and} \quad \eta_\varepsilon = X_\varepsilon' \circ X_\varepsilon^{-1},$$

$$\tau_\varepsilon' = \frac{d\tau_\varepsilon}{d\varepsilon} \quad \text{and} \quad \delta_\varepsilon = \tau_\varepsilon' \circ X_\varepsilon^{-1}$$

and

$$u_{\varepsilon,0}' = \frac{du_{\varepsilon(0)}}{d\varepsilon}.$$

Note that the quantities $\sigma_\varepsilon \circ X_\varepsilon^{-1}$, η_ε and δ_ε are Eulerian quantities whereas X_ε' and τ_ε' are Lagrangian ones.
Then, differentiating $\mathcal{U}$, $\mathcal{T}$ with respect to ε, one finally finds

$$\begin{cases} \mathcal{U}' \circ X_\varepsilon^{-1} = \mathbb{L}(\nabla u_\varepsilon(0))\eta_\varepsilon + \mathbb{L}(u_{\varepsilon,0}') + [\eta_\varepsilon \cdot \nabla, \mathbb{U}](\sigma) + \mathbb{U}(\delta_\varepsilon), & (*) \\ \mathcal{T}' = D_1F(g,\tau)g' + D_2F(g,\tau)\tau', & (**) \\ g' \circ X_\varepsilon^{-1} = \mathbb{L}(\nabla\nabla u_{\varepsilon,0})\eta_\varepsilon + \mathbb{L}(\nabla u_{\varepsilon,0}') + [\eta_\varepsilon \cdot \nabla, \mathbb{G}](\sigma) + \mathbb{G}(\delta_\varepsilon), & (***) \end{cases}$$

where g' is the derivative of g given by (3.1) and $[\cdot,\cdot]$ denotes the usual commutator, in particular

$$[\eta_\varepsilon \cdot \nabla, \mathbb{U}](\sigma) = \eta_\varepsilon(t) \cdot \nabla\mathbb{U}(\sigma) - \mathbb{U}(\eta_\varepsilon(t) \cdot \nabla\sigma).$$

Function spaces

We use the following spaces:

i)

$$C^{\alpha,p} = C^\alpha(\mathbb{R}^n) \cap L^p(\mathbb{R}^n), \quad \text{for } \alpha \in (0,1) \text{ and } p \in (1,\infty),$$

with norm

$$||f||_{\alpha,p} = ||f||_{C^\alpha(\mathbb{R}^n)} + ||f||_{L^p(\mathbb{R}^n)};$$

ii)

$$C^{1+\alpha}(\mathbb{R}^n) \quad \text{for } \alpha \subset (0,1),$$

with norm

$$||f||_{C^{1+\alpha}(\mathbb{R}^n)} = ||f||_{L^\infty(\mathbb{R}^n)} + ||\nabla f||_{C^\alpha(\mathbb{R}^n)};$$

iii)

$$C^{1+\alpha,p} = C^{1+\alpha}(\mathbb{R}^n) \cap W^{1,p}(\mathbb{R}^n), \quad \text{for } \alpha \in (0,1) \text{ and } p \in (1,\infty),$$

with norm

$$||f||_{\alpha,p} = ||f||_{C^{1+\alpha}(\mathbb{R}^n)} + ||f||_{W^{1,p}(\mathbb{R}^n)};$$

iv)

$$L^\infty(0,T;C^{\alpha,p}),\quad \text{for } \alpha\in(0,1) \text{ and } p\in(1,\infty),$$

with norm

$$||f||_{L^\infty(0,T;C^{\alpha,p})} = \sup_{t\in[0,T]} ||f(t)||_{\alpha,p};$$

v)

$$Lip(0,T;C^{\alpha,p}),\quad \text{for } \alpha\in(0,1) \text{ and } p\in(1,\infty),$$

with norm

$$||f||_{Lip(0,T;C^{\alpha,p})} = \sup_{t\neq s;s,t\in[0,T]} \frac{||f(t)-f(s)||_{\alpha,p}}{|t-s|} + ||f||_{L^\infty(0,T;C^{\alpha,p})};$$

vi)

$$C^\beta(0,T;C^{\alpha,p}),\quad \text{for } \alpha,\beta\in(0,1) \text{ and } p\in(1,\infty),$$

with norm

$$||f||_{C^\beta(0,T;C^{\alpha,p})} = \sup_{t\neq s;s,t\in[0,T]} \frac{||f(t)-f(s)||_{\alpha,p}}{|t-s|^\beta} + ||f||_{L^\infty(0,T;C^{\alpha,p})};$$

vii)

$$C^\beta(0,T;C^{\alpha,p}),\quad \text{for } \alpha,\beta\in(0,1) \text{ and } p\in(1,\infty),$$

with norm

$$||f||_{C^\beta(0,T;C^{1+\alpha,p})} = \sup_{t\neq s;s,t\in[0,T]} \frac{||f(t)-f(s)||_{1+\alpha,p}}{|t-s|^\beta} + ||f||_{L^\infty(0,T;C^{1+\alpha,p})}.$$

3.1.3 Results about the Stokes system

We present estimates for the operators and maps we have introduced.

We start with estimates for $\mathbb{U}$, $\mathbb{L}$ and $\mathbb{G}$. The precise estimates are given in Theorems 3.2 and 3.3 below.
These two first theorems give us the information that the map $\mathbb{S}$ is a map into the correct space.
Note that these operators appear also in (*), respectively in (* * *), the expressions we obtained by calculating variational derivatives.

Theorem 3.2. *Let* $0<\alpha<1$, $1<p<\infty$ *and* $T>0$.
Then there exists a constant C *such that*

$$||\mathbb{U}(\sigma)||_{L^\infty(0,T;C^{\alpha,p})} \le C\sqrt{T}||\sigma||_{L^\infty(0,T;C^{\alpha,p})},$$

and

$$||\mathbb{L}(u_0)||_{L^\infty(0,T;C^{\alpha,p})} \le C||u_0||_{\alpha,p}.$$

Theorem 3.3. *For $0 < \alpha < 1$, $0 < \beta \le 1$, $1 < p < \infty$ and $T > 0$, $\mathbb{G}$ is a continuous map between the following spaces*

$$\mathbb{G} : C^{\beta}(0, T; C^{\alpha,p}) \to L^{\infty}(0, T; C^{\alpha,p}).$$

Apart from estimates for the involved operators $\mathbb{U}$, $\mathbb{L}$ and $\mathbb{G}$ we need also suitable estimates for the commutators $[\eta \cdot \nabla, \mathbb{U}]$ and $[\eta \cdot \nabla, \mathbb{G}]$ appearing in $(\star)$ and $(\star\star\star)$, respectively.
Actually, the fact that these terms are commutators is crucial in order to find the desired estimate.

Theorem 3.4. *For $0 < \alpha < 1$, $\frac{1}{2} < \beta \le 1$ and $T > 0$ the following estimate holds*

$$||[\eta \cdot \nabla, \mathbb{G}](\sigma)||_{L^{\infty}(0,T;C^{\alpha,p})} \le C||\eta||_{C^{\beta}(0,T;C^{1+\alpha})}||\sigma||_{C^{\beta}(0,T;C^{\alpha,p})}.$$

Theorem 3.5. *Let $0 < \alpha < 1$, $0 < \beta \le 1$ and $T > 0$. Then we can estimate*

$$||[\eta \cdot \nabla, \mathbb{U}]||_{L^{\infty}(0,T;C^{\alpha,p})} \le C\left(T^{1-\beta}||\eta||_{C^{\beta}(0,T;C^{\alpha})} + T^{1/2}||\eta||_{L^{\infty}(0,T;C^{1+\alpha})}\right)||\sigma||_{L^{\infty}(0,T;C^{\alpha,p})}.$$

These commutator estimates actually are based on the following harmonic-analysis-lemma which is - from a purely technical point of view - the core estimate. Note that the exclusion of $p = 1$ and $p = \infty$ becomes natural once we deal with Calderòn-Zygmund operators (see e.g. Stein's book [15]).

Lemma 3.6. *Assume that $0 < \alpha < 1$ and $1 < p < \infty$. Moreover, let*

$$(\mathbb{K}\sigma)(x) = P.V. \int_{\mathbb{R}^n} k(x - y)\sigma(y)\, dy$$

be a classical Calderòn-Zygmund operator with kernel k which is smooth away from the origin, homogeneous of degree $-n$ and with mean zero on spheres centered at the origin. Then the following estimate holds

$$||[\eta \cdot \nabla, \mathbb{K}]\sigma||_{\alpha,p} \le C||\eta||_{C^{1+\alpha,p}}||\sigma||_{\alpha,p}.$$

Once we have proven estimates for the operators $\mathbb{U}$, $\mathbb{L}$ and $\mathbb{G}$ and for the commutators $[\eta \cdot \nabla, \mathbb{G}]$ and $[\eta \cdot \nabla, \mathbb{U}]$, we can study the map $\mathcal{S}$ given in (3.2).

Theorem 3.7 ($\mathcal{S}$ is a contraction in a suitable space). *Let $0 < \alpha < 1$, $1/2 < \beta < 1$ and $1 < p < \infty$ and assume that $u_0 \in C^{1+\alpha,p}$ and $\sigma_0 \in C^{\alpha,p}$ are given. In addition set*

$$\mathcal{I} = \left\{(X, \tau)\middle|\ ||(X - \mathbb{I}d, \tau)||_{\mathcal{P}_1} \le \Gamma, \frac{1}{2} < |\nabla_a X(a, t)| \le \frac{3}{2}\right\},$$

where Γ is twice the norm of the initial data and

$$\mathcal{P}_1 = Lip(0, T; C^{1+\alpha,p}) \times Lip(0, T; C^{\alpha,p}).$$

Then there exists T sufficiently small such that

$$\mathcal{S} : \mathcal{I} \to \mathcal{I}.$$

Moreover, $\mathcal{S}$ *is a contraction in the space*

$$\mathcal{P} = C^{\beta}(0, T; C^{1+\alpha,p}) \times C^{\beta}(0, T; C^{\alpha,p}),$$

i.e.,

$$||\mathcal{S}(X_1, \tau_1) - \mathcal{S}(X_2, \tau_2)||_{\mathcal{P}} \leq \frac{1}{2}||(X_1 - X_2, \tau_1 - \tau_2)||_{\mathcal{P}}$$

for $(X_1, \tau_1), (X_2, \tau_2) \in \mathcal{I}$.

In a final step, the properties of the map $\mathcal{S}$ can be exploited - together with the earlier established estimates - in order to establish existence and uniqueness of the Stokes systems (*S*) - (S-C).

Theorem 3.8 (Existence and uniqueness for Stokes systems). *Let* $0 < \alpha < 1$ *and* $1 < p < \infty$ *and let a divergence-free* $u_0 \in C^{1+\alpha,p}$ *and* $\sigma_0 \in C^{\alpha,p}$ *be given. Then*

i) there exists T and a solution (u, σ) *of (S) - (S-C) such that* $(u, \sigma) \in L^{\infty}(0, T; C^{1+\alpha,p}) \times Lip(0, T; C^{\alpha,p})$

ii) for two solutions $(u_j, \sigma_j) \in L^{\infty}(0, T; C^{1+\alpha,p}) \times Lip(0, T; C^{\alpha,p})$, $j = 1, 2$ *the corresponding time derivatives of the (particle) path satisfy the Lipschitz bound*

$$||\partial_t X_2 - \partial_t X_1||_{L^{\infty}(0,T;C^{1+\alpha,p})} \leq C(T)\big(||u_2(0) - u_1(0)||_{C^{1+\alpha,p}} + ||\sigma_2 - \sigma_1||_{\alpha,p}\big).$$

In particular, the estimate in ii) implies uniqueness of the solution for given initial data.

All the above mentioned results can be found together with their proofs in [4].

In order to give some indication about the underlying ideas, here we will give a proof of Theorem 3.5.

Proof of Theorem 3.5.

Roughly speaking, the idea is to rewrite the expression we have to estimate in such a manner that we have a commutator under the integral.

First of all, recall that the commutator we want to estimate is nothing else than

$$\begin{aligned}
[\eta \cdot \nabla, \mathbb{U}](\sigma) &= \eta(t) \cdot \nabla \mathbb{U}(\sigma) - \mathbb{U}(\eta(t) \cdot \nabla \sigma) \\
&= \eta \cdot \nabla \left(\int_0^t e^{(t-s)\Delta} \mathbb{H}\, div\, \sigma(s)\, ds \right) - \int_0^t e^{(t-s)\Delta} \mathbb{H}\, div\, (\eta(s) \cdot \nabla \sigma(s))\, ds.
\end{aligned}$$

In this form, we add zero in the following way

$$
\begin{aligned}
[\eta \cdot \nabla, \mathbb{U}](\sigma) &= \eta \cdot \nabla\left(\int_0^t e^{(t-s)\Delta}\mathbb{H}\, div\, \sigma(s)\, ds \right) - \int_0^t e^{(t-s)\Delta}\mathbb{H}\, div\, (\eta(s) \cdot \nabla\sigma(s))\, ds \\
&= \eta \cdot \nabla\left(\int_0^t e^{(t-s)\Delta}\mathbb{H}\, div\, \sigma(s)\, ds \right) - \int_0^t e^{(t-s)\Delta}\mathbb{H}\, div\, (\eta(t) \cdot \nabla\sigma(s))\, ds \\
&\quad - \int_0^t e^{(t-s)\Delta}\mathbb{H}\, div\, (\eta(s) \cdot \nabla\sigma(s)) + \int_0^t e^{(t-s)\Delta}\mathbb{H}\, div\, (\eta(t) \cdot \nabla\sigma(s))\, ds \\
&= \eta \cdot \nabla\left(\int_0^t e^{(t-s)\Delta}\mathbb{H}\, div\, \sigma(s)\, ds \right) - \int_0^t e^{(t-s)\Delta}\mathbb{H}\, div\, (\eta(t) \cdot \nabla\sigma(s))\, ds \\
&\quad - \int_0^t e^{(t-s)\Delta}\mathbb{H}\, div\, ((\eta(s) - \eta(t)) \cdot \nabla\sigma(s))\, ds \\
&= I(t) + II(t).
\end{aligned}
$$

In a next step, we rewrite $II(t)$ as follows

$$
II(t) = -\int_0^t e^{(t-s)\Delta}\mathbb{H}\, div\, ((\eta(s) - \eta(t)) \cdot \nabla\sigma(s))\, ds = E_1(t) + E_2(t),
$$

where

$$
E_1(t) = \mathbb{U}(\nabla \cdot (\eta(s) - \eta(t))\sigma(s)) = \int_0^t e^{(t-s)\Delta}\mathbb{H}\, div\, (\{\nabla \cdot (\eta(s) - \eta(t))\}\sigma(s))\, ds
$$

and

$$
E_2(t) = \int_0^t \nabla\nabla g_{t-s} \star \mathbb{H}((\eta(s) - \eta(t))\sigma(s))\, ds.
$$

In this last expression, g_{t-s} denotes the Gaussian kernel providing the solution of the heat equation.

Thanks to this new formulation we find the estimate

$$
\begin{aligned}
\|II(t)\|_{\alpha,p} &\leq \|E_1(t)\|_{\alpha,p} + \|E_2(t)\|_{\alpha,p} \\
&\leq C\sqrt{t}\|\eta\|_{L^\infty(0,T;C^{1+\alpha})}\|\sigma\|_{L^\infty(0,T;C^{\alpha,p})} + Ct^{1-\beta}\|\eta\|_{C^\beta(0,T;C^\alpha)}\|\sigma\|_{L^\infty(0,T;C^{\alpha,p})}.
\end{aligned}
$$

In the last estimate we used Theorem 3.2 in order to estimate $E_1(t)$ and for the second term we used the properties of the Gaussian kernel g_{t-s}.

Our next goal is to bound

$$I(t) = \eta \cdot \nabla\left(\int_0^t e^{(t-s)\Delta}\mathbb{H}\, div\, \sigma(s)\, ds\right) - \int_0^t e^{(t-s)\Delta}\mathbb{H}\, div\,(\eta(t)\cdot\nabla\sigma(s))\, ds.$$

This term too, we rewrite. Here, we do this in the following way. First of all, note that the first integral can be rewritten as a convolution

$$\begin{aligned} I(t) &= \eta \cdot \nabla\left(\int_0^t g_{t-s} \star \mathbb{H}\, div\, \sigma(s)\, ds\right) - \int_0^t e^{(t-s)\Delta}\mathbb{H}\, div\,(\eta(t)\cdot\nabla\sigma(s))\, ds \\ &= \int_0^t \eta(t)\nabla\nabla g_{t-s} \star \mathbb{H}\,\sigma(s)\, ds - \int_0^t e^{(t-s)\Delta}\mathbb{H}\, div\,(\eta(t)\cdot\nabla\sigma(s))\, ds. \end{aligned}$$

Next, we want to switch the position of the multiplication with $\eta(t)$. This is achieved as follows.

$$\begin{aligned} I(t) &= \int_0^t \eta(t)\nabla\nabla g_{t-s} \star \mathbb{H}\,\sigma(s)\, ds - \int_0^t e^{(t-s)\Delta}\mathbb{H}\, div\,(\eta(t)\cdot\nabla\sigma(s))\, ds \\ &= \int_0^t \eta(t)\nabla\nabla g_{t-s} \star \mathbb{H}\,\sigma(s)\, ds - \int_0^t \nabla\nabla g_{t-s} \star \eta(t)\mathbb{H}\,\sigma(s)\, ds \\ &\quad + \int_0^t \nabla\nabla g_{t-s} \star \eta(t)\mathbb{H}\,\sigma(s)\, ds - \int_0^t e^{(t-s)\Delta}\mathbb{H}\, div\,(\eta(t)\cdot\nabla\sigma(s))\, ds \\ &= D(x,t) + \int_0^t \nabla\nabla g_{t-s} \star \eta(t)\mathbb{H}\,\sigma(s)\, ds - \int_0^t e^{(t-s)\Delta}\mathbb{H}\, div\,(\eta(t)\cdot\nabla\sigma(s))\, ds \\ &= D(x,t) + \int_0^t \nabla g_{t-s} \star\ div\,\big(\eta(t)\mathbb{H}\,\sigma(s)\big)\, ds - \int_0^t g_{t-s} \star \mathbb{H}\, div\,(\eta(t)\cdot\nabla\sigma(s))\, ds \\ &= D(x,t) + \int_0^t \nabla g_{t-s} \star\ div\,\big(\eta(t)\mathbb{H}\,\sigma(s)\big)\, ds - \int_0^t \nabla g_{t-s} \star \mathbb{H}(\eta(t)\cdot\nabla\sigma(s))\, ds \\ &= D(x,t) + \int_0^t \nabla g_{t-s} \star \big(div\,\eta(t)\big)\mathbb{H}\,\sigma(s)\big)\, ds + \int_0^t \nabla g_{t-s} \star \eta(t)\big(div\,\mathbb{H}\,\sigma(s)\big)\, ds \\ &\quad - \int_0^t \nabla g_{t-s} \star \mathbb{H}(\eta(t)\cdot\nabla\sigma(s))\, ds \\ &= D(x,t) + \int_0^t \nabla g_{t-s} \star \big(div\,\eta(t)\big)\mathbb{H}\,\sigma(s)\big)\, ds \end{aligned}$$

$$
\begin{aligned}
&+\int_0^t \nabla g_{t-s} \star \eta(t)\big(div\, \mathbb{H}\, \sigma(s)\big)\, ds - \int_0^t \nabla g_{t-s} \star \mathbb{H}(\eta(t)\cdot \nabla\sigma(s))\, ds \\
= \;& D(x,t) + \int_0^t \nabla e^{(t-s)\Delta}(div\, \eta(t))\mathbb{H}\, \sigma(s)\big)\, ds + \int_0^t \nabla e^{(t-s)\Delta}[\eta(t)\cdot\nabla, \mathbb{H}]\, \sigma(s)\, ds \\
= \;& D(x,t) + III(t) + IV(t).
\end{aligned}
$$

Estimating these expressions leads to

$$
\begin{aligned}
||I(t)||_{\alpha,p} &\le ||D(\cdot,t)||_{\alpha,p} + ||III(t)||_{\alpha,p} + ||IV(t)||_{\alpha,p} \\
&\le C\sqrt{t}||\eta||_{L^\infty(0,T;C^{1+\alpha})}||\sigma||_{L^\infty(0,T;C^{\alpha,p})},
\end{aligned}
$$

where we used again the properties of g_{t-s} and Lemma 3.6.
Putting together all the estimates we have and taking the supremum in time this completes the proof of Theorem 3.5. □

3.1.4 Results about the Navier-Stokes system

Finally, we study the Navier-Stokes systems (NS) - (C).
Since we work in a situation which is subcritical for the Navier-Stokes equations (see hypothesis of the theorems below) we can treat the problem perturbatively.
This is also the reason why we put so much emphasis on the analysis of the related Stokes systems.
Actually, it turns out that the following analogues of Theorems 3.7 and 3.8 hold (cf. Theorems 3.9 and 3.10 below).

Before we give the precise statements, let us say a few words about the necessary modifications in comparison to the case of the Stokes systems we have treated so far.

First of all, note that now, the equations (NS), $\partial_t u + u\cdot\nabla u - \nu\Delta u + \nabla p = div\,(\sigma)$, can be rewritten as

$$\partial_t u - \nu\Delta u + \nabla p = div\,(\sigma - u\otimes u).$$

The formula for the solution u becomes

$$u(x,t) = \mathbb{L}(u_0)(x,t) + \mathbb{U}(\sigma)(x,t) - \mathbb{U}(u\otimes u)(x,t)$$

with gradient

$$\nabla u = \mathbb{L}(\nabla u_0) + \mathbb{G}(\sigma) - \mathbb{U}(\nabla(u\otimes u)),$$

where the operators $\mathbb{L}$, $\mathbb{U}$ und $\mathbb{G}$ are the same as before.

Similarly to the case of the Stokes systems, we introduce the Lagrangian variables X and τ. But here, we will take into consideration also the Lagrangian velocity v given by

$$v = u \circ X.$$

With the notation

$$\mathcal{V} = \mathbb{L}(u_0) \circ X + \mathbb{U}(\tau \circ X^{-1}) \circ X - \mathbb{U}((v \circ v) \circ X^{-1}) \circ X,$$

we have

$$\begin{aligned}\mathcal{V} &= \frac{dX}{dt} \\ &= \mathbb{L}(u_0) \circ X + \mathbb{U}(\tau \circ X^{-1}) \circ X - \mathbb{U}((v \circ v) \circ X^{-1}) \circ X,\end{aligned}$$

where as before

$$\tau = \sigma \circ X$$

and the initial data

$$X(a, 0) = a.$$

And we still have the equation

$$\mathcal{T} = F(g, \tau),$$

where now - the adaptation of (3.1) -

$$g(a, t) = \mathbb{L}(\nabla u_0) \circ X + \mathbb{G}(\tau \circ X^{-1}) \circ X - \mathbb{U}(\nabla(v \otimes v)) \circ X^{-1}) \circ X.$$

With these notations the substitute of the map $\mathcal{S}$ (see (3.2)) is

$$\mathcal{S}^*(X, \tau, v) = \begin{pmatrix} a + \int_0^t \mathcal{V}(X(s, \tau(s), v(s))\, ds \\ \sigma_0(a) + \int_0^t \mathcal{T}(X(s), \tau(s), v(s))\, ds \\ \mathcal{V}(X, \tau, v) \end{pmatrix}.$$

Proceeding with the calculation of variational derivatives and estimating the involved terms carefully, it can be shown that the map $\mathcal{S}^*$ has the suitable properties, in particular it is a contraction if we restrict our attention to a nice function space.

Theorem 3.9. *Let $0 < \alpha < 1$, $1/2 < \beta < 1$ and $1 < p < \infty$ and assume $u_0 \in C^{1+\alpha,p}$ with $\operatorname{div} u_0 = 0$ and $\sigma_0 \in C^{\alpha,p}$ are given. In addition set*

$$\mathcal{I}^* = \left\{ (X, \tau, v) \Big| \; \|(X - \mathbb{I}d, \tau, v)\|_{\mathcal{P}^*_1} \leq \Gamma, \frac{1}{2} < |\nabla_a X(a, t)| \leq \frac{3}{2}, v = \frac{dX}{dt} \right\},$$

where Γ is smaller than the norm of the initial data and where

$$\mathcal{P}^*_1 = Lip(0, T; C^{1+\alpha,p}) \times Lip(0, T; C^{\alpha,p}) \times L^\infty(0, T; C^{1+\alpha,p}).$$

Then there exists T sufficiently small such that

$$\mathcal{S}^* : \mathcal{I}^* \to \mathcal{I}^*.$$

Moreover, there exist $\gamma > 0$ *and* $T > 0$ *such that* $\mathcal{S}^\star$ *is a contraction in the space*

$$\mathcal{P}^\star = C^\beta(0, T; C^{1+\alpha,p}) \times C^\beta(0, T; C^{\alpha,p}) \times \gamma L^\infty(0, T; C^{1+\alpha,p}),$$

i.e.,

$$\begin{aligned} ||\mathcal{S}(X_1, \tau_1, v_1) - \mathcal{S}(X_2, \tau_2, v_2)||_{\mathcal{P}^\star} &\leq \frac{1}{2}||(X_1 - X_2, \tau_1 - \tau_2)||_{\mathcal{P}^\star} \\ &= ||X_1 - X_2||_{C^\beta(0,T;C^{1+\alpha,p})} + ||\tau_1 - \tau_2||_{C^\beta(0,T;C^{\alpha,p})} \\ &\quad + \gamma||v_2 - v_1||_{L^\infty(0,T;C^{1+\alpha,p})} \end{aligned}$$

for $(X_1, \tau_1, v_1), (X_2, \tau_2, v_2) \in \mathcal{I}^\star$.

Finally, we can establish existence and uniqueness also for the Navier-Stokes systems (NS) - (C).

Theorem 3.10 (Existence and uniqueness for Navier-Stokes systems). *Let* $0 < \alpha < 1$ *and* $1 < p < \infty$ *and let a divergence-free* $u_0 \in C^{1+\alpha,p}$ *and* $\sigma_0 \in C^{\alpha,p}$ *be given. Then*

i) there exists T and a solution (u, σ) *of (NS) - (C) such that* $(u, \sigma) \in L^\infty(0, T; C^{1+\alpha,p}) \times Lip(0, T; C^{\alpha,p})$;

ii) for two solutions $(u_j, \sigma_j) \in L^\infty(0, T; C^{1+\alpha,p}) \times Lip(0, T; C^{\alpha,p})$, $j = 1, 2$ *the corresponding time derivatives of the (particle) path satisfy the Lipschitz bound*

$$\begin{aligned} &||\partial_t X_2 - \partial_t X_1||_{L^\infty(0,T;C^{1+\alpha,p})} + ||\partial_t \tau_2 - \partial_t \tau_1||_{L^\infty(0,T;C^{\alpha,p})} \\ &\leq C(T)\big(||u_2(0) - u_1(0)||_{C^{1+\alpha,p}} + ||\sigma_2 - \sigma_1||_{\alpha,p}\big). \end{aligned}$$

In particular, the estimate in ii) implies uniqueness of the solution for given initial data.

The proofs of Theorems 3.9 and 3.10 again can be found in Constantin's article [4].

3.2 Long-time behavior for critical surface quasi geostrophic equation

3.2.1 Introduction

Let us consider the surface quasi geostrophic equation (SQG)

$$\begin{aligned} \partial_t \theta + u \cdot \nabla \theta + \Lambda \theta &= f, \\ \theta(x, \cdot) &= \theta_0, \end{aligned} \tag{3.3}$$

where $\theta = \theta(x, t) \in \mathbb{R}$, $x \in \mathbb{T}^2 = [-\pi, \pi]^2$, $t \geq 0$, $\Lambda = (-\Delta)^{\frac{1}{2}}$ and f is a time-independent force which belongs to $L^\infty(\mathbb{T}^2) \cap H^1(\mathbb{T}^2)$. We assume that the datum and the force have zero mean, i.e., $\int_{\mathbb{T}^2} f(x)\,dx = \int_{\mathbb{T}^2} \theta_0(x)\,dx = 0$, which immediately implies that

$$\int_{\mathbb{T}^2} \theta(x, t)\,dx = 0.$$

The vector u is related to θ by

$$u = \mathcal{R}^{\perp}\theta = (-\mathcal{R}_2\theta, \mathcal{R}_1\theta), \tag{3.4}$$

where $\mathcal{R}_j = \partial_j \Lambda^{-1}$ is the jth Riesz transform.

Remark 3.11. *The choice of working in $\mathbb{T}^2$ rather than in $\mathbb{R}^2$ (where the analytical tools are nicer) is connected to the question of finite dimensionality of the attractors in H^1, which we address in the next sections. We work in two dimensions although many of the results we show can be easily generalized to higher dimensions.*

The term *critical* used for equation (3.3) is related to the fact that the L^∞–norm is scaling invariant. Indeed if θ is a solution of SQG in $\mathbb{T}^2 \times [0, T]$ with initial data θ_0, then also

$$\theta_\lambda(x, t) = \theta\left(\frac{x}{\lambda}, \frac{t}{\lambda}\right) \tag{3.5}$$

is a solution of SQG in $[-\lambda\pi, \lambda\pi]^2 \times [0, \lambda T]$ with initial data $\theta_{0,\lambda}(x) = \theta(x/\lambda)$ and force $f_\lambda(x) = \frac{1}{\lambda} f(x/\lambda)$. Clearly $||\theta||_{L^\infty} = ||\theta_\lambda||_{L^\infty}$. The fractional Laplacian Λ appearing in the equations can be replaced by the "power" fractional Laplacian Λ^s, $s \in \mathbb{R}$. The dynamics generated by dissipative fractional operators with exponents $s < 1$ (*supercritical* case) are of great interest.

The question of *regularity for all times* has been extendedly studied. Here we summarize some important results in this context and we refer to [7] for more details. In the case of $f = 0$ and $\theta_0 = \theta(x, 0) \in C^\infty(\mathbb{T}^2)$, Kiselev, Nazarov and Volberg [13] constructed a family of Lipschitz moduli of continuity that are invariant in time: if the initial data obeys such moduli, then the solution of the critical SQG will also obey them for all later times. In [1], Caffarelli and Vasseur proved the global regularity result in three steps: in a the first step they prove that if u is in L^2, then u is bounded. In a second, fundamental step, they used the De Giorgi iteration method to show that a bounded weak solution instantaneously becomes Hölder continuous for all times. The last step consists into proving that if $u \in C^\alpha$ for any $\alpha > 0$ then $u \in C^\infty$. Kiselev and Nazarov [12] give a different proof of global regularity for SQG using the *molecular duality* method. In 2011 Constantin and Vicol [8] introduced a *non-linear maximum principle* (or rather *non-linear lower bound*) for *linear* nonlocal operators like the fractional Laplacian. The non-linear lower bound for Λ (which immediately proves global regularity of solutions with small L^∞ norm) is the main tool to prove global regularity for the critical SQG only under the additional assumption of *stability under small shocks*: there exists $0 < \delta < 1$ and $L > 0$ such that $|\theta(x, t) - \theta(y, t)| \leq \delta$ whenever $|x - y| \leq L$. In [7] the smallness assumption coming from the "only small shocks" property has been replaced with a smallness condition on the exponent of the Hölder space C^α. The advantage of the method presented in [7] is to get propagation of C^α regularity without appealing to the complex De Giorgi iteration technique. In the same paper the authors proved the existence of a global attractor with a finite Hausdorff dimension. Working in H^1, which is the largest Hilbert space in which uniqueness of weak solutions of SQG is available, the authors prove the existence of a compact absorbing set in phase space. In [9] the authors establish the existence of a compact global attractor for the dynamical system generated by (3.3), which *uniformly* attracts bounded sets in $H^1(\mathbb{T}^2)$. Their proof relies on the combination of the De Giorgi iteration technique and the non-linear maximum principle established in [8].

3.2.1.1 Preliminaries: the fractional Laplacian

We start by defining the fractional Laplacian in $\mathbb{R}^d$: for any $f \in C^{1+\varepsilon}(\mathbb{R}^d)$ for some $\varepsilon > 0$ we can define the fractional Laplacian $\Lambda = (-\Delta)^{\frac{1}{2}}$ in Fourier variables by

$$\mathcal{F}((-\Delta)^{\frac{1}{2}} f)(k) = |k| \mathcal{F} f(k), \qquad k \in \mathbb{R}, \tag{3.6}$$

where $\mathcal{F} f$ is the Fourier transform of f:

$$\mathcal{F} f(k) = \frac{1}{(2\pi)^2} \int f(x)\, e^{-ik\cdot x}\, dx .$$

In real variables

$$(-\Delta)^{\frac{1}{2}} f(x) = \mathcal{F}^{-1}(|k| \mathcal{F} f)(x) = c\mathrm{P.V.} \int_{\mathbb{R}^d} \frac{f(x) - f(y)}{|x-y|^{d+1}}\, dz, \qquad x \in \mathbb{R}^d, \tag{3.7}$$

where the c is a normalizing constant and the integral is understood in the principal value sense.

By (3.4) we can rewrite the vector field u as

$$u = \mathcal{R}^{\perp}\theta = \nabla^{\perp}\Lambda^{-1}\theta = (-\partial_2(-\Delta)^{-\frac{1}{2}}\theta, \partial_1(-\Delta)^{-\frac{1}{2}}\theta) . \tag{3.8}$$

In Fourier variables the relation between u an θ is even more transparent. Indeed (neglecting for a moment the dependency on time) relation (3.8) in Fourier-space transforms into

$$\mathcal{F} u(k) - \frac{ik^{\perp}}{|k|} \mathcal{F}\theta .$$

The fractional Laplacian appears naturally in physical models (see [16]): In the theory of stochastic-processes the Poisson semigroup $e^{-t(-\Delta)^{\frac{1}{2}}}$ is the expectation of a Levy process. The function $v(x,t) := e^{-t(-\Delta)^{\frac{1}{2}}} f(x)$ is the solution of the fractional diffusion equation

$$\begin{cases} \partial_t v = (-\Delta)^{\frac{1}{2}} v, & \text{in } \mathbb{R}^d \times (0,\infty), \\ v(x,0) = f(x), & \text{in } \mathbb{R}^d . \end{cases} \tag{3.9}$$

In elasticity theory, the problem of finding the configuration of and elastic membrane in equilibrium on a thin obstacle motivates the study of the differential equation

$$\begin{cases} \partial_z^2 w + \Delta_y w &= (-\Delta)^{\frac{1}{2}} w, \quad \text{in} \quad (y,z) \in \mathbb{R}^{d-1} \times (0,\infty), \\ w(y,0) &= h(y), \qquad \text{in} \quad y \in \mathbb{R}^{d-1}. \end{cases} \tag{3.10}$$

The solution $w(x, t)$ to this problem is given by the convolution with the Poisson kernel in the upper half space :

$$w(y,z) = e^{-z(-\Delta_y)^{\frac{1}{2}}} h(y) = \int_{\mathbb{R}^d} P(z,(y-\xi))h(\xi)\, d\xi\,,$$

where $P(z,\eta) = c\frac{z}{(z^2+\eta^2)^{\frac{d+1}{2}}}$. Taking the derivative of w with respect to z and estimating it at zero, we realize

$$\partial_z w(y,z)|_{z=0} = -(-\Delta_y)^{\frac{1}{2}} h(y)\,.$$

Notation. *In the rest of the notes we will always abbreviate the fractional Laplacian* $(-\Delta)^{\frac{1}{2}}$ *with the symbol* Λ*.*

If $x \in \mathbb{R}^2$ and θ is a Schwarz function, by definition (3.7) we have

$$\Lambda\theta(x) = c\text{P.V.} \int \frac{\theta(x)-\theta(y)}{|x-y|^3}\, dy, \tag{3.11}$$

where $c > 0$. The following proposition gives us a representation of Λ in the periodic case.

Proposition 3.12 ([10]). *If* $x \in \mathbb{T}^2$ *and* θ *is a Schwarz function, then*

$$\Lambda\theta(x) = c \sum_{v\in\mathbb{Z}^2} \text{P.V.} \int_{\mathbb{T}^2} \frac{\theta(x)-\theta(y)}{|x-y-v|^3}\, dy, \tag{3.12}$$

where $c > 0$.

We recall that Cordoba and Cordoba in [10] proved the following useful pointwise estimate: for $x \in \mathbb{R}^2$ (or $x \in \mathbb{T}^2$) and θ a Schwarz function, we have

$$2\theta\Lambda\theta(x) \geq \Lambda\theta^2(x)\,. \tag{3.13}$$

Indeed, observing that for a generic convex function φ

$$\varphi(f(x)) - \varphi(f(y)) \leq \frac{\varphi'(f(x)) - \varphi'(f(y))}{2}(f(y)-f(x)),$$

and using a symmetrization trick one can easily prove that

$$\Lambda\varphi'(f) \geq \Lambda(\varphi(f))\,. \tag{3.14}$$

In Lemma 3.14 we state a result which is similar to (3.13) but more general and will be a key ingredient for the proof of global regularity.

3.2.1.2 Energy estimates

Testing the equation (3.3) with θ (periodic and mean-zero) and integrating by parts, we get the energy estimate

$$\frac{1}{2}\frac{d}{dt}||\theta||_{L^2}^2 + ||\Lambda^{\frac{1}{2}}\theta||_{L^2}^2 = \int f\theta\, dx\,,$$

where the non-linear term disappeared because (by definition of u) $\nabla \cdot u = 0$.

Let us now suppose $f = 0$ and test the equation (3.3) with $\Delta\theta$ and integrate by parts. We have

$$\frac{1}{2}\frac{d}{dt}||\nabla\theta||_{L^2}^2 - ||\Lambda^{\frac{3}{2}}\theta||_{L^2}^2 = \int (u\cdot\nabla)\theta\Delta\theta\, dx\,. \tag{3.15}$$

Observe that, integrating by parts we have

$$\begin{aligned}
\int (u\cdot\nabla)\theta\Delta\theta &\le \left|\int \partial_i(u_j\partial_j\theta)\partial_i\theta\, dx\right| \\
&= \left|\int \partial_i u_j \partial_j\theta\partial_i\theta\, dx + \int u_j\partial_j\partial_i\theta\partial_i\theta\, dx\right| && (3.16)\\
&\le \int |\nabla u||\nabla\theta|^2\, dx && (3.17)\\
&\le \int |\nabla\theta|^3\, dx && (3.18)\\
&\lesssim \left(\int |\nabla\theta|^2\, dx\right)^{\frac{1}{2}} \left(\int |\Lambda^{\frac{3}{2}}\theta|^2\, dx\right), && (3.19)
\end{aligned}$$

where in (3.16) we used the divergence-free condition for u. To eliminate the second term in (3.18) we used the Calderòn-Zygmund estimate (see (3.8)) and in (3.19) we used the Interpolation Inequality.

Inserting (3.19) into (3.15) we obtain

$$\frac{1}{2}\frac{d}{dt}||\nabla\theta||_{L^2}^2 + (1 - ||\nabla\theta||_{L^2})||\Lambda^{\frac{3}{2}}\theta||_{L^2}^2 \lesssim 0\,,$$

up to universal constants. It is now clear that the solution looses control when $||\nabla\theta||_{L^2} \gtrsim 1$, while it remains controlled if $||\nabla\theta||_{L^2} \lesssim 1$. This is another instance of the criticality of the equation (3.3).

Weak solutions are known to exist for initial data in L^2 ([14]) but uniqueness is only known to be true in the smaller Hilbert space H^1. The existence for short times of solutions of the unforced SQG with initial data in H^1 has been proved in [11] as a stability result of the equation posed in $H^{1+\varepsilon}$. The same result is true for the forced SQG. Observe that the norm $||f||_{\dot{H}^1}^2 = \int_{\mathbb{R}^2} |\xi|^2|\hat{f}(\xi)|^2\, d\xi$ is invariant under the natural scaling (3.5) of SQG. Therefore also the space $H^1(\mathbb{R}^2)$ is critical.

We recall the following well-posedness result which summarizes local in time existence and the global in time regularity (see [9]) for the forced SQG (3.3).

Proposition 3.13 ([9]). *Assume that $f \in L^\infty \cap H^1$. Then, for all initial data $\theta_0 \in H^1$ the initial value problem (3.3) admits a global solution*

$$\theta \in C([0,\infty); H^1) \cap L^2_{\mathrm{loc}}((0,\infty); H^{3/2}).$$

Moreover, θ satisfies the energy inequality

$$\|\theta(t)\|_{L^2}^2 + \int_0^t \|\Lambda^{\frac{1}{2}}\theta(s)\|_{L^2}^2\, ds \le \|\theta_0\|_{L^2}^2 + \frac{1}{c_0}\|f\|_{L^2}^2 t, \qquad \forall t > 0, \tag{3.20}$$

and the decay estimate

$$\|\theta(t)\|_{L^2} \le \|\theta_0\|_{L^2} e^{-c_0 t} + \frac{1}{c_0}\|f\|_{L^2}, \qquad \forall t \ge 0, \tag{3.21}$$

where c_0 is a universal positive constant. If $\theta_0 \in L^\infty$, then we have

$$\|\theta(t)\|_{L^\infty} \le \|\theta_0\|_{L^\infty} e^{-c_0 t} + \frac{1}{c_0}\|f\|_{L^\infty} \qquad \forall t \ge 0. \tag{3.22}$$

Estimate (3.22) follows from an application of the following Lemma (see [5], appendix A) which can be interpreted as an analog of the Poincaré estimate for the fractional Laplacian Λ.

Lemma 3.14 ([5]). *Let $p = 4q$, $q \ge 1$ and suppose $\theta \in C^\infty$ to have zero mean on $\mathbb{T}^2$. Then*

$$\int_{\mathbb{T}^2} \theta^{p-1}(x)\Lambda\theta(x)\, dx \ge \frac{1}{p}\|\Lambda^{\frac{1}{2}}(\theta^{\frac{p}{2}})\|_{L^2}^2 + \frac{\|\theta\|_{L^p}^p}{C}, \tag{3.23}$$

holds, with an explicit constant $C \ge 1$ which is independent of p.

Proof of Proposition 3.13. On the one hand multiplying (3.3) by θ

$$\partial_t\theta\theta + u\cdot\nabla\theta\theta + \theta\Lambda\theta = f\theta,$$

and integrating over $\mathbb{T}^2$ we obtain

$$\int_{\mathbb{T}^2} \frac{1}{2}\frac{d}{dt}|\theta|^2\, dx + \int_{\mathbb{T}^2} u\cdot\nabla\frac{|\theta|^2}{2}\, dx + \int_{\mathbb{T}^2} \theta\Lambda\theta\, dx \le \frac{1}{2}\int_{\mathbb{T}^2} |f|^2\, dx + \frac{1}{2}\int_{\mathbb{T}^2} |\theta|^2\, dx, \tag{3.24}$$

where we applied Young's inequality on the right-hand side. After an integration by parts we use the incompressibility condition for u and apply (3.23) to the right-hand-side of (3.24) to conclude (3.20).

On the other hand, if in (3.24) we apply Lemma 3.14 to the second term of the left-hand side, we get

$$\frac{d}{dt}\|\theta\|_{L^2}^2 + c_0\|\theta\|_{L^2}^2 \le \|f\|_{L^2}^2. \tag{3.25}$$

Solving this ode we find (3.21). With the same argument (multiply (3.3) by θ^{p-1}, integrate by parts and use the incompressibility condition) we obtain the L^p decay estimate

$$||\theta(\cdot,t)||_{L^p} \le ||\theta_0||_{L^p} e^{-tc_0} + \frac{1}{c_0}||f||_{L^p}. \tag{3.26}$$

Since the domain is periodic we can further estimate the right-hand side as

$$||\theta(\cdot,t)||_{L^p} \le (2\pi)^{\frac{2}{p}} ||\theta_0||_{L^\infty} e^{-tc_0} + \frac{(2\pi)^{\frac{2}{p}}}{c_0}||f||_{L^\infty},$$

and, letting $p \to \infty$, we conclude (3.22). □

The bound (3.22) is used in [7] in order to show that the solutions of the forced SQG become bounded in L^∞ with a bound that depends only on the norm of the force and not of the initial data, after a time that depends on the initial H^1 data.

3.2.1.3 The non-linear maximum principle

In this section we introduce the non-linear lower bound for Λ, proved in [8] in the more general case of operators of the form Λ^s with $0 < s < 2$.

Theorem 3.15 ([8]). *Let $\theta \in \mathcal{S}(\mathbb{R}^2)$, where $\mathcal{S}(\mathbb{R}^2)$ is the space of two-dimensional Schwartz functions. For a fixed $i \in \{1, \cdots, d\}$, let $g(x) = \partial_i\theta(x)$. Assume that $\bar{x} \in \mathbb{R}^2$ such that $g(\bar{x}) = \max_{x\in\mathbb{R}^2} g(x) > 0$. Then we have*

$$\Lambda g(\bar{x}) \ge \frac{g(\bar{x})^2}{c||\theta||_{L^\infty}}, \tag{3.27}$$

where the constant c may be computed explicitly.

We observe that the shape-dependent bound (3.27) scales linearly with f but it has a non-linear dependency on the maximum of g.

Proof. Let $r > 0$ be fixed (to be chosen at the end) and let ϕ be a radially, non-decreasing smooth cut-off function which vanishes for $|x| \le 1$ and it is 1 on $|x| \ge 2$, and $|\nabla\phi| \le 4$. We have

$$\begin{aligned}(\Lambda g)(\bar{x}) &\overset{(3.7)}{=} c\int_{\mathbb{R}^2} \frac{g(\dot{x}) - g(\dot{x})}{|x-\bar{x}|^3}\,dx \\ &\ge c\int_{\mathbb{R}^2} \frac{g(\bar{x}) - g(x)}{|x-\bar{x}|^3}\,\phi\left(\frac{x-\bar{x}}{r}\right)dx \\ &\ge c\int_{|x-\bar{x}|\ge 2r} \frac{g(\bar{x})}{|x-\bar{x}|^3}\,dx - c\int_{\mathbb{R}^2} \frac{g(x)}{|x-\bar{x}|^3}\,\phi\left(\frac{x-\bar{x}}{r}\right)dx\end{aligned}$$

$$
\begin{aligned}
&\geq cg(\bar{x}) \int_{|x-\bar{x}|\geq 2r} \frac{1}{|x-\bar{x}|^3}\,dx - \int_{|x-\bar{x}|\geq r} \frac{\partial_i\theta}{|x-\bar{x}|^3}\,\phi\left(\frac{x-\bar{x}}{r}\right) dx \\
&\geq cg(\bar{x}) \int_{|x-\bar{x}|\geq 2r} \frac{1}{|x-\bar{x}|^3}\,dx + \int_{|x-\bar{x}|\geq r} |\theta| \left|\partial_i\left(\frac{1}{|x-\bar{x}|^3}\,\phi\left(\frac{x-\bar{x}}{r}\right)\right)\right| dx \\
&\geq cg(\bar{x}) \int_{|x-\bar{x}|\geq 2r} \frac{1}{|x-\bar{x}|^3}\,dx - ||\theta||_{L^\infty} \int_{|x-\bar{x}|\geq r} \left|\partial_i\left(\frac{1}{|x-\bar{x}|^3}\,\phi\left(\frac{x-\bar{x}}{r}\right)\right)\right| dx \\
&= c_1\frac{g(\bar{x})}{r} - c_2\frac{||\theta||_{L^\infty}}{r^2}\,.
\end{aligned}
\tag{3.28}
$$

Optimizing in r in the last expression we obtain (3.27). □

A similar result is also valid when the function θ is periodic:

Theorem 3.16 ([8]). *Let $\theta \in C^\infty(\mathbb{T}^2)$ and $g = \partial_i f$. Given $\bar{x} \in \mathbb{T}^2$ such that $g(\bar{x}) = \max_{x\in\mathbb{T}^2} g(x)$, there exists a constant c such that either*

$$g(\bar{x}) \leq c||\theta||_{L^\infty}$$

or

$$\Lambda g(\bar{x}) \geq \frac{g(\bar{x})^2}{c||\theta||_{L^\infty}}\,.$$

We recall that any sufficiently regular solution of the equation (3.3) satisfies a maximum principle [7]:

$$||\theta||_{L^\infty} \leq ||\theta_0||_{L^\infty} \tag{3.29}$$

for all $t \geq 0$. The application of the operator $\nabla^\perp$ to equation (3.3) yields

$$\partial_t\nabla^\perp\theta + u\cdot\nabla\nabla^\perp\theta + \Lambda\nabla^\perp\theta = \nabla u\nabla^\perp\theta\,,$$

which in component can be written as

$$\partial_t\partial_i\theta + u_j\partial_j\partial_i\theta + \Lambda\partial_i\theta = \partial_j u_i\partial_i\theta\,.$$

Multiplying the equation above by $\partial_i\theta$, we obtain

$$\frac{1}{2}(\partial_t + u\cdot\nabla)|\nabla\theta|^2 + \partial_i\theta\Lambda\partial_i\theta = -\partial_i u_j\partial_j\theta\partial_i\theta\,.$$

What can we say about $\partial_i\theta\Lambda\partial_i\theta$? In the following theorem we show a useful pointwise lower bound for the term of the product $(g\Lambda g)(x)$.

Theorem 3.17 ([8]). *For $g := \partial_i\theta \in \mathcal{S}(\mathbb{R}^2)$ we have the pointwise bound*

$$(g\Lambda g)(x) \geq \left(\Lambda\frac{g^2}{2}\right)(x) + \frac{|g(x)|^3}{c||\theta||_{L^\infty}}\,. \tag{3.30}$$

Proof.

$$
\begin{aligned}
(g\Lambda g)(x) &\overset{(3.7)}{=} \int_{\mathbb{R}^2} \frac{g(x)(g(x)-g(y))}{|x-y|^3}\,dy \\
&= \frac{c}{2}\int_{\mathbb{R}^2} \frac{(g(x)+g(y))(g(x)-g(y))}{|x-y|^{d+1}} + \frac{c}{2}\int_{\mathbb{R}^d} \frac{|g(x)-g(y)|^2}{|x-y|^3}\,dy \\
&= \frac{1}{2}\Lambda g^2 + \frac{c}{2}\int_{\mathbb{R}^3} \frac{|g(x)-g(y)|^2}{|x-y|^3}\,dy \\
&\geq \frac{1}{2}\Lambda g^2 + c_1\frac{g^2(x)}{r} - 2g(x)\int_{\mathbb{R}^2} \frac{\partial_y\theta}{|x-y|^3}\,dy \\
&\geq \frac{1}{2}\Lambda g^2 + c_1\frac{g^2(x)}{r} - 2g(x)\int_{|x-y|\geq r} \frac{\partial_y\theta}{|x-y|^3}\,dy \\
&\geq \frac{1}{2}\Lambda g^2 + \frac{g^2(x)}{r} - 2c\frac{|g(x)|\,||\theta||_{L^\infty}}{r^2} \\
&\geq \frac{1}{2}\Lambda g^2 + \frac{g(x)}{r}\left[c_1|g(x)| - \frac{c_2}{r}||\theta||_{L^\infty}\right],
\end{aligned}
$$

where in the fourth line we argued as in (3.28)[3.1]. Optimizing in r we obtain

$$
(g\Lambda g)(x) \geq \frac{\Lambda g^2}{2} + c\frac{|g|^3}{||\theta||_{L^\infty}}.
$$

□

3.2.2 Main Results: global regularity

Our goal is to show global regularity: starting with an initial data in H^1, we will show that the solution to (3.3) gains regularity in time. The proof consists of two steps: using the De Giorgi iteration technique combined with a-priori estimates we show that starting with an initial data in H^1, the solution of (3.3) becomes bounded. We show that the L^∞ bound depends only on $||f||_{L^\infty}$ and on the time it takes to enter the absorbing ball (bounded set $B \in H^1$). In a second stage, exploiting the non-linear maximum principle (3.27), we show that the bounded solutions gains C^α–regularity. The main difficulties in the proof of the global regularity are the *criticality*, the *quasi-linearity* of the equation and the elimination of the initial-data dependency from the final estimates.

3.1 Note that the rigorous argument requires the use of cut-off functions.

Note that in his PhD thesis Resnick (1994) shows the existence of a global weak solutions in L^2. By De Giorgi's iteration method and using the inequality (3.14) one can show that a global weak solution in L^2 is also in L^∞.

3.2.2.1 From L^2 to L^∞ via De Giorgi iteration technique

Theorem 3.18. *Let B be a bounded set in $H^1(\mathbb{T}^2)$ and let θ be a solution of (3.3) with initial data $\theta_0 \in H^1(\mathbb{T}^2)$. Then there exists a universal constant c_0 and a time $t_B \geq 0$ (depending only on B and f) such that*

$$||\theta(t)||_{L^\infty(\mathbb{T}^2)} \leq \frac{1}{c_0}||f||_{L^\infty(\mathbb{T}^2)}, \tag{3.31}$$

for all $t \geq t_B$.

Remark 3.19. *Note that the bound is independent of the L^2 norm of the initial data.*

Proof. The proof consists of two steps: we first show that starting with an initial data in L^2, the solution of SQG becomes bounded (see (3.37)). In a second step we show that, starting with an initial data in H^1, we obtain an L^∞– bound of the solution of SQG, depending only on the L^∞–norm of f and the time it takes to become bounded (see (3.31)).

Let $\theta(\tau) \in L^\infty$ for almost every $\tau \in (\frac{1}{2}, 1)$. For $M \gg 1$ (to be fixed later) we denote by M_k the levels

$$M_k = M(1 - 2^{-k}),$$

and by θ_k the truncated function

$$\theta_k(t) = (\theta(t) - M_k)_+ = \max\{\theta(t) - M_k, 0\}.$$

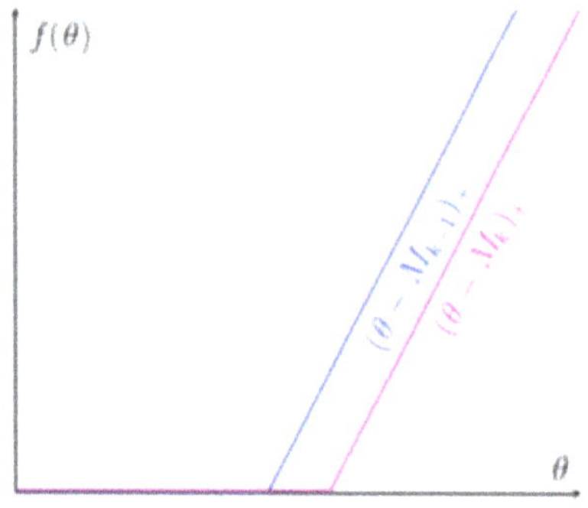

We observe that

(obs 1) θ_k is convex,
(obs 2) $\theta_k \leq \theta_{k-1}$,
(obs 3) Since $\theta_{k-1} \geq 2^{-k}M$ on $\{(x,t) : \theta_k(x,t) > 0\}$, then $\mathbb{1}_{\{\theta_k>0\}} \leq \frac{2^k}{M}\theta_{k-1}$.

Let us define the cut-off function

$$\tau_k = \frac{1}{2}(1 - 2^{-k}).$$

By the energy inequality (see eq. (3.15)) we have

$$\begin{aligned}\frac{d}{dt}||\theta_k||_{L^2}^2 + ||\Lambda^{\frac{1}{2}}\theta_k||_{L^2}^2 &\leq \int_{\mathbb{T}} |f||\theta_k|\,dx \\ &\leq ||f||_{L^\infty}||\theta_k||_{L^1}.\end{aligned}$$

Integrating in time, $\tau \in (s,t)$, we have

$$||\theta_k(t)||_{L^2}^2 + 2\int_s^t ||\Lambda^{\frac{1}{2}}\theta_k(\tau)||_{L^2}^2\,d\tau \leq ||\theta_k(s)||_{L^2}^2 + 2||f||_{L^\infty}\int_s^t ||\theta_k(\tau)||_{L^1}\,d\tau.$$

Taking $s \in (\tau_{k-1}, \tau_k)$ and $t \in (\tau_k, 1]$ we have

$$\sup_{\tau\in[t_k,1]} ||\theta_k(t)||_{L^2}^2 + 2\int_{\tau_k}^1 ||\Lambda^{\frac{1}{2}}\theta_k(\tau)||_{L^2}^2\,d\tau \leq ||\theta_k(s)||_{L^2}^2 + 2||f||_{L^\infty}\int_{\tau_{k-1}}^1 ||\theta_k(\tau)||_{L^1}\,d\tau.$$

Finally, averaging over $s \in (\tau_{k-1}, \tau_k)$ we find

$$\sup_{\tau\in[t_k,1]} ||\theta_k(t)||_{L^2}^2 + 2\int_{\tau_k}^1 ||\Lambda^{\frac{1}{2}}\theta_k(\tau)||_{L^2}^2\,d\tau \tag{3.32}$$

$$2^k\int_{\tau_{k-1}}^1 ||\theta_k(s)||_{L^2}^2\,ds + 2||f||_{L^\infty}\int_{\tau_{k-1}}^1 ||\theta_k(\tau)||_{L^1}\,d\tau. \tag{3.33}$$

We define the quantity

$$Q_k = \sup_{\tau\subset[t_k,1]} ||\theta_k(t)||_{L^2}^2 + 2\int_{\tau_k}^1 ||\Lambda^{\frac{1}{2}}\theta_k(\tau)||_{L^2}^2\,d\tau, \tag{3.34}$$

and observe that, thanks to (3.20), we can bound Q_0 with the norms of θ_0 and f

$$Q_0 \leq ||\theta_0||_{L^2}^2 + \frac{1}{c_0}||f||_{L^2}^2. \tag{3.35}$$

We claim that

$$Q_k \lesssim \frac{2^{2k}}{k}Q_{k-1}^{\frac{3}{2}}, \tag{3.36}$$

and that

$$||\theta(t)||_{L^\infty} \le c\,[||\theta_0||_{L^2} + ||f||_{L^2}]\, e^{-c_0 t} + \frac{1}{c_0}||f||_{L^\infty} . \tag{3.37}$$

It is easy to see that (3.36) implies (3.37): indeed, if we ensure

$$M \ge c\sqrt{Q_0}\,, \tag{3.38}$$

then the sequence Q_k converges to zero as $k \to \infty$. By (3.20) we have

$$||(\theta(t) - M)_+||_{L^2} = 0\,,$$

and (applying the same argument above to $(\theta(t) - M)_-$)

$$||(\theta(t) - M)_-||_{L^2} = 0\,,$$

from which we deduce

$$|\theta| \le M \qquad \text{for a.e } \tau \in (1/2, 1)\,. \tag{3.39}$$

Since the condition (3.38) is in particular satisfied when

$$M \ge c\,[||\theta_0||_{L^2} + ||f||_{L^2}]\,,$$

passing to the infimum in M in (3.39) we obtain

$$||\theta||_{L^\infty} \le c\,[||\theta_0||_{L^2} + ||f||_{L^2}]\,, \qquad \text{for a.e } \tau \in (1/2, 1)\,.$$

Finally, from the decay estimate (3.22) we deduce the uniform bound (3.37) for all $t > 0$, for some constant $c > 0$.

We now show how to conclude (3.31) starting from (3.37). Let us fix a bounded set $B \subset H^1$ and let

$$R = ||B||_{H^1} = \sup_{\phi \in B} ||\phi||_{H^1}\,.$$

From (3.37) and the Poincaré inequality we deduce that if $\theta_0 \in B$ then

$$||\theta(t)||_{L^\infty} \le c\,[R + ||f||_{L^2}]\, e^{-c_0 t} + \frac{1}{c_0}||f||_{L^\infty} \qquad \text{for all } t \ge 1\,.$$

Define the entering time $t_B = t_B(R, ||f||_{L^2 \cap L^\infty}) \ge 1$ so that for all $t \ge t_B$

$$[R + ||f||_{L^2}]\, e^{-c_0 t} \le \frac{1}{c_0}||f||_{L^\infty}\,,$$

from which we can conclude (3.31).

In what follows we sketch the argument for (3.36); for more details the reader can consult [9].

Recall that by the definition (3.34) and (3.32) we have

$$Q_k \le 2^k \int_{\tau_{k-1}}^{1} ||\theta_k(s)||_{L^2}^2\, ds + 2||f||_{L^\infty} \int_{\tau_{k-1}}^{1} ||\theta_k||_{L^1}(\tau)\, d\tau\,.$$

The bound (3.36) is a consequence of the crucial estimate

$$||\theta_k||^2_{L^3(\mathbb{T}^2\times[\tau_k,1])} \le cQ_k, \qquad \text{for all } k \in \mathbb{N}. \tag{3.40}$$

Estimate (3.40) follows from

$$||\theta||^2_{L^3} \lesssim ||\Lambda^{\frac{1}{2}}\theta||^{\frac{4}{3}}_{L^2}||\theta||^{\frac{2}{3}}_{L^2},$$

which, in turn, is a combination of the Hölder estimate

$$||\theta||^2_{L^3} \lesssim ||\theta||^{\frac{4}{3}}_{L^2}||\theta||^{\frac{2}{3}}_{L^2},$$

and the Sobolev-embedding estimate ($H^{\frac{1}{2}} \subset L^4$)

$$||\theta||_{L^4(\mathbb{T}^2)} \lesssim ||\Lambda^{\frac{1}{2}}\theta||_{L^2(\mathbb{T}^2)}.$$

The crucial estimate (3.40) and the observations (obs2) and (obs3) above yield

$$\begin{aligned} 2^k\int_{\tau_{k-1}}^{1} ||\theta_k(s)||^2_{L^2} &\le c\frac{2^{2k}}{M}Q^{\frac{3}{2}}_{k-1}, \\ \int_{\tau_{k-1}}^{1} ||\theta_k||_{L^1} &\le c\frac{2^{2k}}{M^2}Q^{\frac{3}{2}}_{k-1}. \end{aligned}$$

□

3.2.2.2 From L^∞ to C^α via non-linear maximum principle

Theorem 3.20. *Let $\theta_0(t_1) \in L^\infty(\mathbb{T}^2)$. Then there exist $t_2 = t_2(f) > t_1$ and $\alpha = \alpha(||\theta_0||_{L^\infty}, ||f||_{L^\infty}) > 0$ such that*

$$||\theta(t_2)||_{C^\alpha(\mathbb{T}^2)} \le c||f||_{L^\infty(\mathbb{T}^2)} \tag{3.41}$$

where $||\theta||_{C^\alpha} = ||\theta||_{L^\infty} + [\theta]_{C^\alpha}$ and $[\theta]_{C^\alpha} = \sup_{x\neq y\in\mathbb{T}^2} |\phi(x) - \phi(y)||x - y|^{-\alpha}$.

In order to estimate C^α–seminorms, it is natural to consider the finite difference $\delta_h\theta(x, t) = \theta(x + h, t) - \theta(x, t)$, which is periodic in both x and h, where $x, h \in \mathbb{T}^2$. Note that $\delta_h\theta$ satisfies (see argument for (3.27))

$$(\delta_h\theta\Lambda\delta_h\theta)(x) = \left(\Lambda\frac{(\delta_h\theta)^2}{2}\right)(x) + D(\delta_h\theta),$$

where $D(\delta_h\theta) = \int \frac{(\delta_h\theta(x)-\delta_h\theta(y))^2}{|x-y|^3}\,dy$. By the same argument used for (3.27), the term $D(\delta_h\theta)$ can be easily estimated: [3.2]

3.2 The rigorous argument requires the use of cut-off function as in (3.27). For details of the computation we refer to [7], Theorem 4.3.

$$
\begin{aligned}
D(\delta_h\theta)(x) &\geq \int_{\mathbb{R}^2} \frac{(\delta_h\theta(x)-\delta_h\theta(y))^2}{|x-y|^3}\,dy \\
&\geq \int_{|x-y|\geq\rho} \frac{(\delta_h\theta(x)-\delta_h\theta(y))^2}{|x-y|^3}\,dy \\
&\geq c_1\frac{|\delta_h\theta(x)|^2}{\rho} - c|\delta_h\theta(x)| \int_{|x-y|\geq\rho} \frac{\delta_h\theta(y)}{|x-y|^3}\,dy \\
&\geq c_1\frac{|\delta_h\theta(x)|^2}{\rho} - c_2|\delta_h\theta(x)|\frac{|h|}{\rho^2}||\theta||_{L^\infty}\,.
\end{aligned}
$$

Optimizing the right hand side of the last expression in ρ we obtain

$$D(\delta_h\theta) \geq C\frac{|\delta_h\theta|^3}{|h|||\theta||_{L^\infty}}, \tag{3.42}$$

and we can conclude

$$(\delta_h\theta\Lambda\delta_h\theta)(x) \geq \Lambda\frac{(\delta_h\theta)^2}{2} + c_1\frac{|\delta_h\theta|^3}{|h|||\theta||_{L^\infty}}\,. \tag{3.43}$$

Remark 3.21. *Before sketching the proof of Theorem 3.20, we illustrate the argument in a simpler situation. Let us consider the unforced Burgers's equation*

$$\partial_t\theta + \theta\theta_x + \Lambda\theta = 0\,. \tag{3.44}$$

Multiplying (3.44) by $\frac{\delta_h\theta}{|h|^{2\alpha}}$ *we obtain*

$$\frac{1}{2}(\partial_t + \theta\partial_x + \delta_h\theta\partial_h + \Lambda)\frac{(\delta_h\theta)^2}{|h|^{2\alpha}} + c_1\frac{|\delta_h\theta|^3}{|h|||\theta||_{L^\infty}}\frac{1}{|h|^{2\alpha}} \leq c_2\alpha\frac{|\delta_h\theta|^3}{|h|^{2\alpha+1}}\,.$$

We choose α *small enough such that*

$$\frac{c}{||\theta||_{L^\infty}} \geq c_2\alpha \qquad \Rightarrow \qquad \alpha||\theta||_{L^\infty} \ll 1$$

and we can absorb the non-linear term of the right hand side in the left hand side of the identity above.

Sketch of the proof of Theorem 3.20. (For the detailed proof the reader may refer to [7], Theorem 4.3.)
Let us consider the forced surface quasi geostrophic equation

$$\partial_t\theta + u\cdot\nabla\theta + \Lambda\theta = f, \qquad \text{for } f \in L^\infty \tag{3.45}$$

and assume $\theta_0 = \theta(x, t_0) \in L^\infty$ (bounded in terms of $\|f\|_{L^\infty}$). The finite difference $\delta_h\theta = \theta(x+h) - \theta(x)$ satisfies the equation

$$\partial_t\delta_h\theta + \delta_h(u\cdot\nabla\theta) + \Lambda\delta_h\theta = \delta_h f\,. \tag{3.46}$$

We can express the middle term of the left-hand-side as

$$\begin{aligned}\delta_h(u\cdot\nabla\theta) &= u(x+h)\nabla\theta(x+h) - u(x)\nabla\theta(x)\\ &= u(x)\nabla(\theta(x+h)-\theta(x)) + (\delta_h u)\nabla_x\theta(x+h)\\ &= (u\cdot\nabla)\delta_h\theta + (\delta_h u)\cdot\nabla_x(\delta_h\theta)\,,\end{aligned}$$

and rewrite (3.46) as

$$\partial_t\delta_h\theta + u\cdot\nabla_x\delta_h\theta + (\delta_h u)\nabla_h\delta_h\theta + \Lambda\delta_h\theta = \delta_h f\,.$$

Multiplying the equation above by $\delta_h\theta$ and using the identity $(v\cdot\nabla\eta)\eta = v\nabla\frac{\eta^2}{2}$ we get

$$\frac{1}{2}(\partial_t + u\cdot\nabla_x + (\delta_h u)\cdot\nabla_h + \Lambda)|\delta_h\theta|^2 + D(h) = (\delta_h f)\delta_h\theta\,, \tag{3.47}$$

where

$$D(h) = c\int\limits_{|x-y|\geq\rho}\frac{((\partial_h\theta)(x) - (\partial_h\theta)(y))^2}{|x-y|^3}\,dy.$$

Dividing (3.47) by $\frac{1}{|h|^{2\alpha}}$

$$\frac{1}{2}(\partial_t + u\cdot\nabla_x + (\delta_h u)\cdot\nabla_h + \Lambda)\left(\frac{|\delta_h\theta|^2}{|h|^{2\alpha}}\right) \tag{3.48}$$

$$+\ \frac{D(h)}{|h|^{2\alpha}} + \frac{2\alpha}{|h|^{2\alpha+1}}\left((\delta_h u)\frac{h}{|h|}\frac{|\delta_h\theta|^2}{2}\right) = \frac{(\delta_h f)\delta_h\theta}{|h|^{2\alpha}}\,, \tag{3.49}$$

and calling the operator $(\partial_t + u\cdot\nabla_x + (\delta_h u)\cdot\nabla_h + \Lambda) =: L$, we have the bound

$$\frac{1}{2}L\left(\frac{|\delta_h\theta|^2}{|h|^{2\alpha}}\right) + \frac{D(h)}{|h|^{2\alpha}} \leq \frac{\|f\|_{L^\infty}|\delta_h\theta|}{|h|^{2\alpha}} + c_3\frac{\alpha|\delta_h u|}{|h|^{2\alpha+1}}|\delta_h\theta|^2\,. \tag{3.50}$$

Observe that, by the relation $u = \mathcal{R}^\perp\theta$ and the definition (3.7), $\delta_h u$ can be written as

$$\delta_h u = \mathcal{R}^\perp\delta_h\theta = \mathrm{P.V.}\int\limits_{\mathbb{R}^d}\frac{(x-y)^\perp}{|x-y|^3}((\delta_h\theta)(y))\,dy\,.$$

It is convenient to *softly* split (i.e., with smooth cut-off functions) $\delta_h u$ into the near field and the far field contributions:

$$\delta_h u = (\delta_h u)_{\text{in}} + (\delta_h u)_{\text{out}}, \qquad \text{where}$$

$$(\delta_h u)_{\text{in}}(x) = \mathcal{R}^\perp\delta_h\theta = \mathrm{P.V.}\int\limits_{|x-y|\leq\rho}\frac{(x-y)^\perp}{|x-y|^3}((\delta_h\theta)(y) - \delta_n\theta(x))\,dy \tag{3.51}$$

$$(\delta_h u)_{\text{out}}(x) \;=\; \mathcal{R}^{\perp}\delta_h\theta = \text{P.V.} \int\limits_{|x-y|>\rho} \frac{(x-y)^{\perp}}{|x-y|^3}((\delta_h\theta)(y))\,dy\,,$$

where in (3.51) we have used that the kernel of $\mathcal{R}^{\perp}$ has zero average in the sphere. The far-field can be easily be estimated by the Hölder inequality as follows:

$$|(\delta_h u)_{\text{out}}| \le \|\theta\|_{L^\infty}|h| \int\limits_{|x-y|>\rho} \frac{1}{|x-y|^{d+1}}\,dy = \|\theta\|_{L^\infty}\frac{|h|}{\rho}\,.$$

To bound the near-field, instead, we use the Cauchy-Schwarz inequality

$$\begin{aligned}|(\delta_h u)_{\text{in}}| &\le \int\limits_{|x-y|\le\rho} \frac{|\delta_h\theta(x)-\delta_h\theta(y)|}{|x-y|^3}\,dy\\ &\le \left(\int\limits_{|x-y|\le\rho} \frac{|\delta_h\theta(x)-\delta_h\theta(y)|^2}{|x-y|^4}\,dy\right)^{\frac{1}{2}} \left(\int\limits_{|x-y|\le\rho} \frac{1}{|x-y|^2}\,dy\right)^{\frac{1}{2}} \le c\sqrt{D_h\rho}\,.\end{aligned}$$

Inserting the bounds for $|\delta_h u_{\text{out}}|$ and $|\delta_h u_{\text{in}}|$ into (3.50) we have

$$\begin{aligned}&\frac{1}{2}L\left(\frac{|\delta_h\theta|^2}{|h|^{2\alpha}}\right) + \frac{D(h)}{|h|^{2\alpha}}\\ &\le c_4\left(\sqrt{2(h)\rho}\right)\frac{\alpha}{|h|}\frac{1}{|h|^{2\alpha}}|\delta_n\theta|^2 + c_5\frac{\alpha}{|h|}\frac{|h|}{\rho}\|\theta\|_{L^\infty}\frac{|\delta_h\theta|^2}{|h|^{2\alpha}} + \frac{\|f\|_{L^\infty}|\delta_h\theta|}{|h|^{2\alpha}}\,.\end{aligned}$$

For simplicity, we now let $f=0$. By Young's inequality

$$\left(\sqrt{D(h)\rho}\right)\frac{\alpha}{|h|}\frac{1}{|h|^{2\alpha}}|\delta_n\theta|^2 \le \varepsilon D(h)\frac{1}{|h|^{2\alpha}} + \frac{\alpha^2}{\varepsilon}\rho\frac{|\delta_h\theta|^4}{|h|^{2\alpha}}\,,$$

and the non-linear bound (3.42), we obtain

$$\begin{aligned}\frac{1}{2}L\left(\frac{|\delta_h\theta|^2}{|h|^{2\alpha}}\right) + \frac{D(h)}{|h|^{2\alpha}} &\overset{(3.42)}{\lesssim} \frac{1}{2}L\left(\frac{|\delta_h\theta|^2}{|h|^{2\alpha}}\right) + \frac{|\delta_h\theta|^3}{|h|^{2\alpha+1}\|\theta\|_{L^\infty}}\\ &\lesssim \rho\frac{\alpha^2}{|h|^2}\frac{|\delta_h\theta|^4}{|h|^{2\alpha}} + \frac{\alpha}{|h|}\frac{|h|}{\rho}\|\theta\|_{L^\infty}\frac{|\delta_h\theta|^2}{|h|^{2\alpha}}\\ &\lesssim \frac{\alpha}{|h|^{2\alpha}}\left[\alpha\rho\frac{|\delta_h\theta|^2}{|h|^2} + \frac{\|\theta\|_{L^\infty}}{\rho}\right]|\delta_h\theta|^2\,,\end{aligned}$$

where the symbol $\lesssim$ stands for $\le$ up to universal constants.
The optimal value of the right-hand side is reached when $\frac{\rho}{|h|} = \sqrt{\frac{\|\theta\|_{L^\infty}}{\alpha|\delta_h\theta|^2}}$. This choice implies

$$\frac{1}{2}L\left(\frac{|\delta_h\theta|^2}{|h|^{2\alpha}}\right) + \frac{|\delta_h\theta|^3}{|h|^{2\alpha+1}\|\theta\|_{L^\infty}} \lesssim \frac{\alpha^{\frac{3}{2}}}{|h|^{2\alpha+1}}\sqrt{\|\theta\|_{L^\infty}}|\delta_h\theta|^3\,, \tag{3.52}$$

and, when $\alpha||\theta||_{L^\infty} \ll 1$, we can conclude

$$\frac{1}{2}L\left(\frac{|\delta_h\theta|^2}{|h|^{2\alpha}}\right) \le 0\,,$$

which implies the C^α-bound (3.41) (see [7] and [9] for details). □

Acknowledgment: The authors would like to thank the organizers, Anna Mazzucato and Gianluca Crippa, for the very interesting Summer School in Levico Terme and the Fondazione Bruno Kessler und the CIRM for their administrative support and hospitality.
The second and third author would like to thank the organizers in addition for having given them the opportunity to report the present lecture notes.

Bibliography

[1] L. Caffarelli and A. Vasseur. Drift diffusion equations with fractional diffusion and the quasi-geostrophic equation. *Annals of Mathematics*, 171(3):1903–1930, 2010.
[2] P. Constantin. An Eulerian-Lagrangian approach for incompressible fluids: Local theory. *Journal of the American Mathematical Society*, 14:263–278, 2001.
[3] P. Constantin. An Eulerian-Lagrangian approach to the Navier-Stokes equations. *Comm. Math. Phys.*, 216, 2001.
[4] P. Constantin. Lagrangian–Eulerian methods for uniqueness in hydrodynamic systems. *Adv. Math.*, 278:67–102, 2015.
[5] P. Constantin, N. Glatt-Holtz, and V. Vicol. Unique Ergodicity for Fractionally Dissipated, Stochastically Forced 2D Euler Equations. *Communications in Mathematical Physics*, 330(2):819–857, 2014.
[6] P. Constantin and W. Sun. Remarks on Oldroyd-B and related complex fluid models. *Commun. Math. Sci.*, 10:33–73, 2012.
[7] P. Constantin, A. Tarfulea, and V. Vicol. Long Time Dynamics of Forced Critical SQG. *Communications in Mathematical Physics*, 335(1):93–141, 2014.
[8] P. Constantin and V. Vicol. Nonlinear maximum principles for dissipative linear nonlocal operators and applications. *Geometric and Functional Analysis*, 22(5):1289–1321, 2012.
[9] P. Constantin, M. C. Zelati, and V. Vicol. Uniformly attracting limit sets for the critically dissipative SQG equation.
[10] A. Córdoba and D. Córdoba. A Maximum Principle Applied to Quasi-Geostrophic Equations. *Communications in Mathematical Physics*, 249(3):511–528, 2004.
[11] N. Ju. Dissipative 2D quasi-geostrophic equation: local well-posedness, global regularity and similarity solutions. *Indiana University mathematics journal*, 56(1):187–206, 2007.
[12] A. Kiselev and F. Nazarov. Global regularity for the critical dispersive dissipative surface quasi-geostrophic equation. *arXiv preprint arXiv:0908.0925*, 2009.
[13] A. Kiselev, F. Nazarov, and A. Volberg. Global well-posedness for the critical 2D dissipative quasi-geostrophic equation. *Inventiones mathematicae*, 167(3):445–453, 2006.
[14] S. Resnick. *Dynamical problems in Non-linear Advances Partial Differential Equations*. PhD thesis, Ph. D. thesis, University of Chicago, Il, 1995.
[15] E. M. Stein. *Singular Integrals and Differentiability Properties of Functions*. Princeton University Press, 1971.

[16] P. R. Stinga. Fractional powers of second order partial differential operators: extension problem and regularity theory. 2010.
[17] M. E. Taylor. *Introduction to Differential Equations*. American Mathematical Society, 2011.

Alexander Kiselev*, Mikhail Chernobay, Omar Lazar, and Chao Li

Small Scale Creation in Inviscid Fluids

Abstract: These lectures overview recent progress in understanding of small scale and finite time singularity formation in solutions of the incompressible Euler equation and related models. We start with the background information on existence and uniqueness of solutions in two dimensions, and proceed to discuss examples of solutions of the 2D Euler equation with fast growth in gradient of vorticity. We then discuss Hou-Luo scenario for finite time blow up in solutions of the 3D Euler equation and simplified 1D models designed to gain insight into some features of the scenario.

Keywords: ideal fluid, incompressible Euler equation, Boussinesq system, axisymmetric flow, regularity of solutions, generation of small scales, singularity formation

4.1 Introduction

Attempts to understand the nature of fluid motion have occupied minds of researchers for many centuries. Fluids are all around us, and we can witness the complexity and subtleness of their properties in every day life, in ubiquitous technology, and in dramatic phenomena such as tornado or hurricane. There has been an enormous wealth of knowledge accumulated in the broad area of fluid mechanics, yet it is quite remarkable that many of the most fundamental as well as important in applications questions remain poorly understood. A special role in fluid mechanics is played by incompressible Euler equation, first formulated in 1755 [10], the second partial differential equation (PDE) ever derived (the first one is wave equation derived by D'Alembert 8 years earlier). It describes motion of an inviscid, volume preserving fluid. The incompressible Euler equation is a non-linear and nonlocal system of PDE, with dynamics near a given point depending on the flow field over the entire region filled with fluid. This makes analysis of these equations exceedingly challenging, and the array of mathematical methods applied to their study has been extremely broad.

The main function of a PDE such as Euler equation is, given some initial data, to allow us to compute the solution and use it for prediction of future behavior of the fluid. This is exactly how weather forecasting works, or how new airplane shapes are designed. Therefore, the first basic question one can ask about a PDE is existence and uniqueness of solutions in some appropriate class. If one can show existence and

***Corresponding Author: Alexander Kiselev:** Department of Mathematics, Duke University, U.S.A., E-mail: kiselev@math.duke.edu
Mikhail Chernobay: St. Petersburg State University, Russian Federation
Omar Lazar: Instituto de Ciencias Mathematicas, Spain, E-mail: omar.lazar@icmat.es
Chao Li: Department of Mathematics, Rice University, U.S.A.

uniqueness of solutions, the PDE is often called globally regular (sometimes, continuous dependence on the initial data is also added to the list of desirable properties). On the other hand, if solutions can form singularities in finite time, the terminology for such phenomena is that finite time blow up happens. Understanding singularities is important because they correspond to dramatic, highly intense fluid motion, can indicate limits of applicability of the model, and are very difficult to resolve computationally. The story of global regularity vs finite time blow up for incompressible Euler equation is very different in two and three spatial dimensions. While for $d = 2$, global regularity is known since 1930s, the question remains open for $d = 3$, where only local existence of regular solutions is known. The reasons for such disparity will become clear below, once we write the Euler equation in vorticity form. More generally, one can ask a related and broader question about creation of small scales in fluids - coherent structures that vary sharply in space and time, and contribute to phenomena such as turbulence. The problem has a long history of contributions by leading mathematicians; mathematically, one often asks about lower bounds on the growth of derivatives of solutions in certain scenarios. One can consult the books [16], [17] for more details on history of the problem.

The main goal of these lectures is to review some recent developments in the area. A few years ago, based on extensive numerical simulations, Hou and Luo [15] have proposed a new scenario for singularity formation in the 3D Euler equation. The scenario has a fascinating, complex geometry - it is axi-symmetric, and growth in vorticity is observed at a ring of hyperbolic points of the flow located at the boundary of a cylinder. The solution has self-similar features which, however, do not appear to be exact. Inspired by this work, Kiselev and Sverak provided a rigorous construction of an example of solutions of 2D Euler equation where growth of vorticity gradient is very fast - double exponential in time. Such rate of growth is known to be sharp. Also, a couple of new one-dimensional models have been developed to gain insight into the Hou-Luo scenario. We will review these and related earlier works and outline some open questions and directions.

Our starting point is a global regularity result for solutions of 2D Euler equation. We will follow the approach by Yudovich, often referred to as Yudovich theory. It establishes existence and uniqueness of solutions for initial data with bounded vorticity, $\omega = \mathrm{curl} u$. As we will see, this class is very natural since it leads to log-Lipschitz fluid velocities ensuring uniqueness of fluid particle trajectories. We will also see that the results can be easily upgraded to more regular initial data and solutions.

We are mostly interested in the study of the Euler equation in a bounded domain $D \subset \mathbb{R}^d$ that is compact and smooth. In the first sections, we will consider the case $d = 2$, and in the later ones discuss one-dimensional models of the three-dimensional phenomena. The incompressible Euler equation reads as follows

$$(\mathcal{E}) : \begin{cases} \partial_t u + (u \cdot \nabla u) = \nabla p, \\ u \cdot n|_{\partial D} = 0, \\ \nabla \cdot u = 0, \\ u(x, 0) = u_0(x). \end{cases} \tag{4.1}$$

It is well known that, when written in terms of the vorticity ω = curl u the Euler equation becomes the following 2D transport equation

$$\partial_t \omega + (u \cdot \nabla)\omega = 0. \tag{4.2}$$

In three dimensions, there is also the vortex stretching term $(\omega \cdot \nabla)u$ on the right hand side, but it vanishes in 2D. The velocity field u is given by the Biot-Savart law $u = \nabla^{\perp}(-\Delta)^{-1}\omega$ where $\nabla^{\perp} = (\partial_2, -\partial_1)$.

On can consider the equation in terms of trajectories $\Phi_t(x)$ (the flow map corresponding to the 2D Euler) that is

$$(\mathcal{E}) : \begin{cases} \frac{d\Phi_t}{dt} = u(\Phi_t(x), t), \\ \omega(\Phi_t(x), t) = \omega_0(x), \\ \Phi_0(x) = x. \end{cases} \tag{4.3}$$

We have

$$\begin{aligned} \left| \frac{d}{dt} |\Phi_t(x) - \Phi_t(y)|^2 \right| &\leq 2 \left| u(\Phi_t(x, t)) - u(\Phi_t(y), t)) \cdot (\Phi_t(x) - \Phi_t(y)) \right| \\ &\leq 2 \|\nabla u(\cdot, t)\|_{L^\infty} |\Phi_t(x) - \Phi_t(y)|^2. \end{aligned}$$

Then, using Grönwall's lemma,

$$\exp\left(- \int_0^t \|\nabla u(\cdot, s)\|_{L^\infty} \, ds \right) \leq \frac{|\Phi_t(x) - \Phi_t(y)|}{|x - y|} \leq \exp\left(\int_0^t \|\nabla u(\cdot, s)\|_{L^\infty} \, ds \right). \tag{4.4}$$

From (4.3), we can infer that $\omega(x, t) = \omega_0(\Phi_t^{-1}(x))$. We are going to assume $\omega_0 \in L^\infty$. We will see that this is a natural class for the existence and uniqueness theory. Let us introduce the notation

$$u = \nabla^{\perp}(-\Delta_D)^{-1}\omega = \nabla^{\perp} \int_D G_D(x, y)\omega(y) \, dy = \int_D \underbrace{\nabla^{\perp} G_D(x, y)}_{K_D(x,y)} \omega(y) \, dy,$$

where

$$G_D(x, y) = \frac{1}{2\pi} \ln(|x - y|) + h(x, y),$$

and h is such that

$$\Delta_x h(x, y) = 0 \text{ and } h(x, y)|_{x \in \partial D} = -\frac{1}{2\pi} \ln(|x - y|).$$

Lemma 4.1. *The estimates stated below are well known:*

$$\begin{aligned} |G_D(x,y)| &\le C(1+\ln(|x-y|)), \\ |\nabla G_D(x,y)| &\le C(D)|x-y|^{-1}, \\ |\nabla^2 G_D(x,y)| &\le C(D)|x-y|^{-2}. \end{aligned}$$

We have the following proposition.

Proposition 4.2. *The following estimate holds: for every $x, x' \in D$,*

$$\int_D |K_D(x,y) - K_D(x',y)|\, dy \le C\rho(|x-x'|),$$

where ρ is defined by $\rho(r) = r(1-\ln(r))$ if $r \le 1$ and $\rho(r) = 1$ if $r \ge 1$.

Sketch of the proof: The main interesting regime is when x and x' are close enough that is $|x-x'| = \delta < 1$. Then we have

$$(i) \quad \int_{D\cap B_{2\delta}(x)} |K_D(x,y) - K_D(x',y)|\, dy \le \int_{B_{3\delta}(x)} \frac{C}{|x-y|}\, dy = C\int_0^{3\delta} \frac{1}{r} r\, dr = c\delta,$$

and

$$\begin{aligned} (ii) \quad \int_{D\cap B^c_{2\delta}(x)} |K_D(x,y) - K_D(x',y)| dy &\le C\delta \int_{D\cap B^c_{2\delta}(x)} |\nabla K_D(x''(y), y)|\, dy \\ &\le \delta \int_\delta^C \frac{r}{r^2}\, dr \le c\delta(1-\ln(\delta)). \end{aligned}$$

Note that due to presence of the boundary, $x''(y)$ may in general not lie on an interval between x and x'; one has instead to use a path which lies entirely in D. We leave details of the argument to interested reader. □

Proposition 4.3. *Assume $u(x,t)$ is log-Lipschitz*

$$|u(x,t) - u(x',t)| \le c\rho(|x-x'|).$$

Then the Cauchy problem

$$(\mathfrak{C}) : \begin{cases} \frac{d}{dt}x(t) = u(x(t),t), \\ x(0) = x_0, \end{cases} \tag{4.5}$$

has a unique solution.

The uniqueness can be proved in the usual way. That is, assume we have two different solutions $x(t)$ and $y(t)$. Set $z(t) = x(t) - y(t)$, then

$$\left|\frac{d}{dt}z\right| \leq |u(x(t), t) - u(y(t), t)| \leq Cz(t)(1 - \ln z(t)),$$

and so

$$\int_{z(0)}^{z(t)} \frac{dz}{z(1 - \ln(z))} \leq Ct.$$

This is impossible if $z(0) = 0$ and $z(t) > 0$ for some t.

To construct a solution, we consider the sequence of equations

$$u_n(x, t) = K_D \star \omega_{n-1}(x, t).$$

It is equivalent to solve the trajectories equation

$$\frac{d\Phi_t^n(x, t)}{dt} = u_n(\Phi_t^n, t)$$

and set $\omega_n(x, t) = \omega_0((\Phi_t^n)^{-1}(x))$. Then, since the sequence ω_n is uniformly bounded it implies that u_n is uniformly log-Lipschitz and we have

$$\left|\frac{d}{dt}|\Phi_t^n(x) - \Phi_t^n(y)|^2\right| \leq C\left|\Phi_t^n(x) - \Phi_t^n(y)\right|^2 \left(1 - \log|\Phi_t^n(x) - \Phi_t^n(y)|\right),$$

(while $|\Phi_t^n(x) - \Phi_t^n(y)| \leq 1/2$) and so

$$\left|\frac{d}{dt}\log|\Phi_t^n(x) - \Phi_t^n(y)|\right| \leq -C\log\left|\Phi_t^n(x) - \Phi_t^n(y)\right|.$$

Grönwall lemma then gives the following two-sided estimate

$$|x - y|^{e^{Ct}} \leq |\Phi_t^n(x) - \Phi_t^n(y)| \leq |x - y|^{e^{-Ct}}.$$

Therefore, Φ_t is uniformly Hölder continuous in x for every t. The same applies to the inverse map Φ_t^{-1} :

$$|x - y|^{e^{Ct}} \leq |(\Phi_t^n)^{-1}(x) - (\Phi_t^n)^{-1}(y) \leq |x - y|^{e^{-Ct}}.$$

Then, on any interval $[0, T] \times \bar{D}$ we can find (via Ascoli-Arzela criterion) a subsequence n_j such that

$$\Phi_t^{n_j}(x) \longrightarrow \Phi_t(x) \in \mathcal{C}^{\alpha(T)}([0, T] \times D),$$

with ω, u defined by $\omega(x, t) = \omega_0(\Phi_t^{-1}(x))$ and $u(x, t) = K_D \star \omega(x, t)$.

Next we state two useful related theorems. The first one summarizes our discussion, while the second one is a natural extension.

A Yudovich's Theorem. *([20],[16],[17]) Assume $\omega_0 \in L^\infty(\bar{D})$ for a bounded domain D. Then there exists a unique triple $(\Phi_t(x), \omega, u)$ satisfying*

$$\omega(x,t) = \omega_0(\Phi_t^{-1}(x)), \quad \frac{d\Phi_t(x)}{dt} = u(\Phi_t(x), t), \quad \text{and } u = K_D \star \omega(x,t).$$

Moreover, u is log-Lipschitz in x uniformly in time and

$$\Phi_t^{-1}(x) \text{ and } \Phi_t(x) \text{ belong to } \mathcal{C}^{\alpha(T)}([0,T] \times \bar{D}), \alpha(T) > 0,$$

and $\omega \in L^\infty$, $\omega(x,t)$ converges to ω_0 as $t \to 0$ in the weak-$\star$ topology in L^∞.

Theorem 4.4. *Suppose in addition that $\omega_0 \in \mathcal{C}^k(\bar{D})$. Then $\omega(x,t) \in \mathcal{C}^k(\bar{D})$, $u \in C^{k,\alpha}$ for all $\alpha < 1$ and its k order derivatives are log-Lipschitz in x, and $\Phi_t(x) \in \mathcal{C}^{k,\alpha(T)}([0,T] \times \bar{D})$.*

Examples of interesting questions are for instance: How fast can the derivative of a solution grow? How fast small scales can be generated? We begin by giving some examples of dynamics, illustrating Yudovich theorem and its sharpness as well as providing first insight into the growth questions. The so-called Bahouri-Chemin [1] example is defined on the torus $\mathbb{T}^2$. This solution has some symmetry; namely it is odd with respect to both coordinate axes x_1 and x_2. Suppose that $\mathbb{T}^2 = [-\pi, \pi)^2$. The vorticity in the Bahouri-Chemin example is identically equal to 1 in the first quadrant $(0,\pi)\times(0,\pi)$ and is defined elsewhere by symmetry and periodicity.

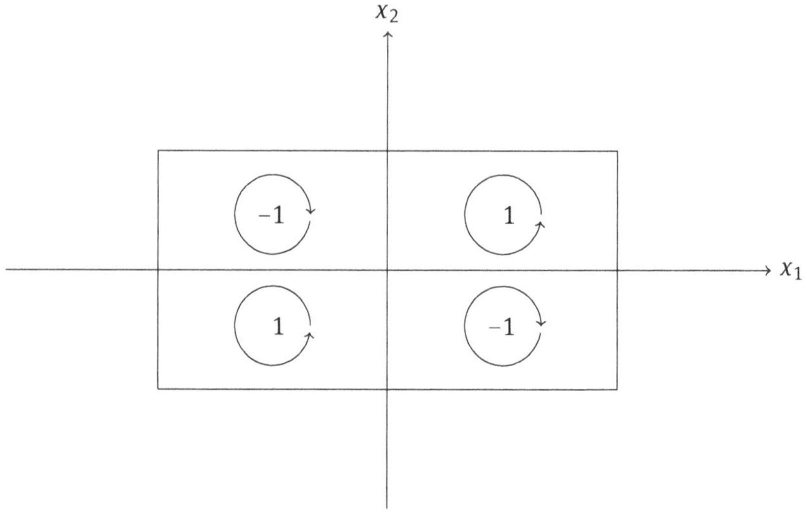

Claim: Such $\omega_0(x)$ is a stationary solution in the sense of Yudovich. The result follows from the lemma below.

Lemma 4.5. *Suppose that the domain D is symmetric with respect to x_2 axis. If $\omega_0 \in L^\infty$ is odd with respect to x_1 then the corresponding solution $\omega(x,t)$ remains odd for all $t > 0$.*

Moreover, the solution corresponding to Bahouri-Chemin example remains constant in time.

Remark 4.6. *Of course, oddness with respect to any other symmetry axis of D is also conserved. The argument also applies to $\mathbb{R}^2$ or $\mathbb{T}^2$.*

Proof. One can check directly that if $\omega(x_1, x_2, t)$ is a solution then so is $-\omega(-x_1, x_2, t)$. At time 0, they are both equal to $\omega_0(x)$ and therefore by uniqueness they coincide for all time. Furthemore, if $\omega(x_1, x_2, t)$ is odd for all time then $(-\Delta)^{-1}\omega(x_1, x_2, t)$ is also odd (this can be checked using Fourier transform on the torus for instance), and $u_1 = \partial_2(-\Delta)^{-1}\omega$ implies oddness of u_1 with respect to x_1. A similar argument applies to u_2 and at other cell boundaries. Thus, all trajectories stay inside the cell where they started for all time. The formula $\omega(x,t) = \omega_0(\Phi_t^{-1}(x))$ then shows that the Bahouri-Chemin solution is stationary. □

For the property of Bahouri-Chemin solution that we want to derive we will need the following lemma.

Lemma 4.7. *Suppose $\omega_0 \in L^\infty(\mathbb{T}^2)$ with mean zero, $u(x) = \nabla^\perp(-\Delta)^{-1}\omega$. Then,*

$$u(x) = \lim_{\gamma \to 0} \frac{1}{2\pi} \int_{\mathbb{R}^2} \frac{(x-y)^\perp}{|x-y|^2} \exp^{-\gamma|y|^2} \omega(y)\, dy,$$

where ω is extended periodically to all $\mathbb{R}^2$.

The proof of this lemma can be established using Fourier transform. We leave details to interested reader.

Theorem 4.8. *In the Bahouri-Chemin example we have*

$$u_1(x_1, 0) = \frac{2}{\pi} x_1 \ln(x_1) + O(x_1)$$

for small x_1.

Proof. We decompose $u_1(x_1, 0) = u_1^M(x_1, 0) + u_1^R(x_1, 0)$, with

$$u_1^M(x_1, 0) \equiv \frac{1}{2\pi} \int_{\mathbb{T}^2} \frac{-y_2}{|x-y|^2} \omega(y)\, dy,$$

and

$$u_1^R(x_1, 0) \equiv \frac{1}{2\pi} \lim_{\gamma \to 0} \int_{\mathbb{R}^2 \setminus \mathbb{T}^2} \frac{-y_2}{|x-y|^2} \omega(y) e^{-\gamma|y|^2}\, dy.$$

Then, using the oddness of $\omega(y)$, we get

$$u_1^R(x_1, 0) = -\frac{1}{\pi} \lim_{\gamma \to 0} \int_{\mathbb{R}^2 \setminus \mathbb{T}^2, y_1 \ge 0, y_2 \ge 0} \left(\frac{y_2}{(x_1 - y_1)^2 + y_2^2} - \frac{y_2}{(x_1 + y_1)^2 + y_2^2} \right) \omega(y)\, dy,$$

hence an estimate using periodicity and mean zero property of ω shows

$$u_1^R(x_1,0) = \frac{1}{\pi}\lim_{\gamma\to 0} \int\limits_{\mathbb{R}^2\setminus\mathbb{T}^2, y_1\geq 0, y_2\geq 0} \frac{4x_1y_1y_2}{((x_1-y_1)^2+y_2^2)((x_1+y_1)^2+y_2^2)} w(y)e^{-\gamma|y|^2}\,dx\,dy \leq Cx_1.$$

The log part will come from u_1^M: indeed, we have

$$u_1^M(x_1,0) = \frac{1}{\pi}\int\limits_0^{\pi}\int\limits_0^{\pi} \frac{4x_1y_1y_2}{((x_1-y_1)^2+y_2^2)(x_1+y_1)^2+y_2^2)} dy_1 dy_2.$$

The main contribution comes from

$$B := \frac{4}{\pi}x_1 \int\limits_{2x_1}^{1}\int\limits_{2x_1}^{1} \frac{y_1y_2}{(y_1^2+y_2^2)^2} dy_1 dy_2.$$

The difference between $u_1^M(x_1,0)$ and B can be estimated as Lipschitz, $\leq Cx_1$. On the other hand, computing the value of B directly gives $B = \frac{2}{\pi}x_1\ln(x_1) + O(x_1)$. □

Another feature of the Bahouri-Chemin example is that the flow map $\Phi_t(x)$ is Hölder with exponent decaying in time. Indeed $d/dt(\Phi_t^1(x_1,0)) \sim c\Phi_t^1(x_1,0)\log(\Phi_t^1(x_1,0))$ (the characteristic on the separatrix). Therefore, we have $d/dt(\log(\Phi_t^1(x_1,0))) \sim c\log(\Phi_t^1(x_1,0))$ so that $\log(\Phi_t^1(x_1,0)) \sim \log(x_1)e^{ct}$. Hence, the inverse flow map $\Phi_t^{-1}(x)$ has Hölder exponent less or equal to e^{-ct}. This means that the estimates of Yudovich theorem are qualitatively sharp.

4.2 An upper bound for growth of $\nabla\omega$

Last lecture, we showed that if $\omega_0 \in L^\infty \Rightarrow \exists$ unique solution (ω, Φ_t, u) of the 2D Euler equation in the following sense:

$$\frac{d\Phi_t}{dt}(x) = u(\Phi_t(x),t), \quad \Phi_0(x) = x, \tag{4.6}$$

$$u(x,t) = \int\limits_D K_D(x,y)\omega(y,t)\,dy, \tag{4.7}$$

$$\omega(x,t) = \omega_0(\Phi_t^{-1}(x)). \tag{4.8}$$

Here u is log-Lipschitz in x, ω is in L^∞, $\Phi_t(x)$ and $\Phi_t^{-1}(x)$ are in $C^{\alpha(T)}([0,T]\times\bar{D})$.

If in addition $\omega_0 \in C^1$, then we have $\omega \in C^{\alpha(T)} \Rightarrow u \in C^{1,\alpha(T)}$. This statement is directly implied by the following classical theorem.

Theorem 4.9 (Kellogg, Schauder, see e.g. [11]). *Suppose that D is a domain in R^d with smooth boundary. Let ψ solve the Dirichlet problem $-\Delta\psi = \omega$ and $\psi|_{\partial D} = 0$. Assume that $\omega \in C^\alpha(\bar{D})$, $\alpha > 0$. Then $\psi \in C^{2,\alpha}(\bar{D})$ and $\|\partial_{ij}\psi\|_{C^\alpha} \leq C(\alpha,D)\|\omega\|_{C^\alpha}$.*

Let us recall the equation (4.4) that we derived last time:

$$\exp\left(-\int_0^t \|\nabla u\|_{L^\infty}\,ds\right) \le \frac{|\Phi_t(x)-\Phi_t(y)|}{|x-y|} \le \exp\left(\int_0^t \|\nabla u\|_{L^\infty(s)}\,ds\right).$$

If $u \in C^{1,\alpha(T)}$, it implies that $\Phi_t(x)$, $\Phi_t^{-1}(x)$ is Lipschitz in x for every t. Moreover, with slightly stronger technical effort one can show that $\Phi_t(x), \Phi_t^{-1}(x) \in C^{1,\alpha(T)}$. This implies that $\omega \in C^1(\bar{D})$ for all times. The next theorem provides a quantitative version of this argument.

Theorem 4.10 (Wolibner, Hölder, Yudovich [19],[12],[20]). *Assume $\omega_0 \in C^1(\bar{D})$, $D \subset R^2$, is compact with smooth boundary. Then the gradient of the solution $\omega(x,t)$ satisfies the following bound*

$$\frac{\|\nabla\omega(\cdot,t)\|_{L^\infty}}{\|\omega_0\|_{L^\infty}} \le \left(\frac{\|\nabla\omega_0\|_{L^\infty}}{\|\omega_0\|_{L^\infty}}+1\right)^{\exp(C\|\omega_0\|_{L^\infty}t)} \tag{4.9}$$

for all $t > 0$.

Ingredients of the proof:

1. Due to the two-sided nature of (4.4), we have

$$e^{-\int_0^t \|\nabla u(\cdot,s)\|_{L^\infty}ds} \le \frac{|\Phi_t^{-1}(x)-\Phi_t^{-1}(y)|}{|x-y|} \le e^{\int_0^t \|\nabla u(\cdot,s)\|_{L^\infty}ds}. \tag{4.10}$$

2. Notice that

$$\begin{aligned}\|\nabla\omega(\cdot,t)\|_{L^\infty} &\le \sup_{x,y\in\bar{D}} \frac{|\omega_0(\Phi_t^{-1}(x))-\omega_0(\Phi_t^{-1}(y)|)}{|x-y|}\\ &\le \|\nabla\omega_0\|_{L^\infty} \sup_{x,y\in\bar{D}} \frac{|\Phi_t^{-1}(x)-\Phi_t^{-1}(y)|}{|x-y|}.\end{aligned}$$

3. Kato's inequality, which we will prove later. If $\omega \in C^\alpha(\bar{D})$, $\alpha > 0$, $u = \nabla^\perp(-\Delta_D)^{-1}\omega$. Then

$$\|\nabla u\|_{L^\infty} \le C(\alpha,D)\|\omega_0\|_{L^\infty}\left(1+\log\left(1+\frac{\|\omega\|_{C^\alpha}}{\|\omega\|_{L^\infty}}\right)\right). \tag{4.11}$$

Combining equations (4.10) and (4.11), we obtain

$$f(t)^{-1} \le \frac{|\Phi_t^{-1}(x)-\Phi_t^{-1}(y)|}{|x-y|} \le f(t), \tag{4.12}$$

where

$$f(t) = \exp\left(C\|\omega_0\|_{L^\infty}\int_0^t \left(1+\log\left(1+\frac{\|\nabla\omega(x,s)\|_{L^\infty}}{\|\omega_0\|_{L^\infty}}\right)\right)ds\right).$$

Combining this together with ingredient 2, we have

$$\log \|\nabla\omega(\cdot,t)\|_{L^\infty} \le \log \|\nabla\omega_0\|_{L^\infty} + C\|\omega_0\|_{L^\infty} \int_0^t (1+\log(1+\frac{\|\nabla\omega(x,s)\|_{L^\infty}}{\|\omega_0\|_{L^\infty}}))ds. \quad (4.13)$$

Let $z = \frac{\|\nabla\omega(\cdot,t)\|_{L^\infty}}{\|\omega_0\|_{L^\infty}}$. Then it is straightforward to show that $z(t) \le y(t)$ where $y(t)$ solves

$$\frac{y'}{y} = C\|\omega_0\|_{L^\infty}(1+\log(1+y)), \qquad y(0) = \frac{\|\nabla\omega_0\|_{L^\infty}}{\|\omega_0\|_{L^\infty}}. \quad (4.14)$$

Hence

$$\frac{y'}{1+y} \le C\|\omega_0\|_{L^\infty}(1+\log(1+y)). \quad (4.15)$$

After integrating both sides from 0 to t, we obtain

$$1+\log(1+y(t)) \le (1+\log(1+y(0))\exp(C\|\omega_0\|_{L^\infty}t). \quad (4.16)$$

□

Proof of Kato's inequality. Take $\delta = \min\{(\frac{\|\omega_0\|_{L^\infty}}{\|\omega_0\|_{C^\alpha}})^{\frac{1}{\alpha}}, \gamma\}$, where γ is chosen so that the set of $x \in D$ with $dist(x, \partial D) \ge 2\delta$ is not empty. According to Biot-Savart law and properties of Dirichlet Green's function, we know

$$u(x,t) = \int K_D(x,y)\omega(y,t)dy, \qquad K_D = \nabla^\perp G_D,$$

$$\nabla u(x,t) = \frac{1}{2\pi}P.V.\int \nabla K_D(x,y)\omega(y,t)dy + M\omega(x),$$

where M is a constant matrix. Note that we need to exercise care when taking derivative of u since singularity in the kernel becomes non-integrable. Computing the derivative in weak sense leads to the extra term $M\omega(x)$ which is of no concern in the estimate we need. First we consider $x \in D$, s.t $dist(x, \partial D) \ge 2\delta$. The part of integral over the complement of the ball centered at x with radius δ can be estimated as

$$\left|\int_{B_\delta^c(x)} \nabla|K_D(x,y)\omega(y,t)dy\right| \le C\|\omega_0\|_{L^\infty} \int_{B_\delta^c(x)} |x-y|^{-2}dy \le C\|\omega_0\|_{L^\infty}(1+\log\delta^{-1}). \quad (4.17)$$

For the other part of the integral, we recall the decomposition of G_D. The first term is

$$
\begin{aligned}
|P.V.\textstyle\int_{B_\delta(x)} \partial^2_{x_i,x_j} \log|x-y|\omega(y)dy| &= |\textstyle\int_{B_\delta(x)} \partial^2_{x_i,x_j} \log|x-y|(\omega(y)-\omega(x))dy| \\
&\le C\|\omega(x,t)\|_{C^\alpha} \textstyle\int_0^\delta r^{-1+\alpha}dr \\
&\le C(\alpha)\delta^\alpha \|\omega(x,t)\|_{C^\alpha} \\
&\le C(\alpha)\|\omega_0\|_{L^\infty}.
\end{aligned}
$$

The last inequality comes from our choice of δ.

Finally, we deal with the last term. Notice that by maximal principle, $|h(z,y)| \le C\log\delta^{-1}$ for all $y \in B_\delta(x)$, and $z \in D$. Then standard estimates for harmonic functions (see e.g. [9]) give, for each $y \in B_\delta(x)$,

$$|\partial^2_{x_i x_j} h(x,y)| \le C\delta^{-4} \int_{B_\delta(x)} |h(z,y)|dz \le C\delta^{-2}\log\delta^{-1}.$$

This gives

$$\left| \int_{B_\delta(x)} \partial^2_{x_i x_j} h(x,y)\omega(y,t)dy \right| \le C\|\omega_0\|_{L^\infty}\log\delta^{-1}. \tag{4.18}$$

Combining these estimates, the inequality is proved for interior points satisfying $dist(x,\partial D) \ge 2\delta$.

Now if x' is such that $dist(x',\partial D) < 2\delta$, find a point x such that $dist(x,\partial D) \ge 2\delta$ and $|x-x'| \le C(D)\delta$. By Theorem 4.9, we have

$$|\nabla u(x') - \nabla u(x)| \le C(\alpha,D)\delta^\alpha\|\omega\|_{C^\alpha}, \tag{4.19}$$

which, together with estimate for interior point x, implies that the inequality holds for all points in D. □

4.3 Simple examples of gradient growth in passive scalars

The passive scalar equation in $2D$ is given by

$$\partial_t\psi + (u\cdot\nabla)\psi = 0, \qquad \psi(x,0) = \psi_0(x).$$

Here u is a given, "passive" vector field that may or may not depend on time.

1. Shear flow

$$u = (u(x_2), 0),$$

$$\Phi_t^{-1}(x_1,x_2) = (x_1 - u(x_2)t, x_2),$$

$$\psi(x,t) = \psi_0(\Phi_t^{-1}(x)).$$

In this example,

$$\|\nabla\psi\|_{L^\infty} \sim ct, \tag{4.20}$$

provided that u' is bounded.

2. Cellular flow

$$\omega(x_1, x_2) = \sin x_1 \sin x_2,$$

$$u(x_1, x_2) = (-\sin x_1 \cos x_2, \sin x_2 \cos x_1),$$

$$\frac{d}{dt}\Phi_t^1(x_1, 0) \sim -\Phi_t^1(x_1, 0) \quad for\ x_2 = 0, \quad x_1\ small.$$

So $x_1(t) \sim x_1(0)e^{-t}$, $\|\nabla\psi\|_{L^\infty} \sim e^{ct}$. We also know if $\|\nabla u\|_{L^\infty} \leq C$, then exponential growth is the fastest that one can get.

3. Bahouri-Chemin flow In this example, described in the first lecture, the flow u satisfies $u_1(x_1, 0) \sim cx_1 \ln x_1$ for x_1 small enough, so $\Phi_t^1(x_1, 0) \sim x_1^{e^{ct}}$ if x_1 is sufficiently small. This leads to double exponential growth of $\|\nabla\psi\|_{L^\infty}$ in a passive scalar advected by such u.

4.4 Growth of derivatives in solutions of 2D Euler

The first works constructing examples with growth in derivatives of vorticity are due to Yudovich [13, 21], He used Lyapunov functional method to prove some growth in $\|\nabla\omega\|_{L^\infty}$ at a flat part of the boundary of domain D. Then, Nadirashvili [18] has constructed examples of the 2D Euler solutions on an annulus with linear growth of $\|\nabla\omega\|_{L^\infty}$. Later Denisov [7] constructed an example in periodic setting that shows superlinear growth; to be specific, he proved $\frac{1}{T^2}\int_0^T \|\nabla\omega(\cdot, t)\|_{L^\infty} dt \to \infty$ as $T \to \infty$. Denisov also built an example to show that $\|\nabla\omega(\cdot, t)\|_{L^\infty}$ can experience bursts of double exponential growth over finite time intervals [8]. The idea for the latter example involves smoothing and slightly perturbing Bahouri-Chemin flow.

We are going to describe a very recent example showing that double exponential growth in the derivatives of solutions of 2D Euler equation in a bounded domain can indeed happen for all times. Thus the upper bound going back to Wolibner is qualitatively sharp.

Theorem 4.11 (Kiselev,Sverak [14]). *Let D be a unit disk in $\mathbb{R}^2$, then there exists $\omega_0 \in C^\infty(\bar{D})$ with $\|\nabla\omega_0\|_{L^\infty} \geq 1$ such that $\|\nabla\omega(\cdot, t)\|_{L^\infty} \geq \|\nabla\omega_0\|_{L^\infty}^{c\exp(ct)}$, for any t.*

As we will see, growth happens at the boundary ∂D. The example is motivated by Luo-Hou's numerical experiments [15], where a new scenario for finite time singularity formation in solutions of the 3D Euler equation is proposed. The scenario is axisymmetric, and extremely fast growth is observed at a ring of hyperbolic points of the flow located at the boundary of a cylinder. We will see below that the geometry of the double exponential growth example is similar, and a hyperbolic point on the boundary plays a key role.

It will be convenient for us to set the origin at the lowest point of the unit disk D

(so that the center of the disk has coordinates (0, 1)). Denote $D^+ = \{x \in D | x_1 \geq 0\}$. The initial data ω_0 will be odd in x_1. Then the solution $\omega(x, t)$ is also odd for all times. By Biot-Savart law, we have

$$u(x, t) = \nabla^\perp \int G_D(x, y)\omega(y, t)dy,$$

where due to our choice of coordinates $G_D(x, y) = \frac{1}{2\pi} \ln \frac{|x-y|}{|x-\bar{y}||y-e_2|}$, $\bar{y} = \frac{y-e_2}{|y-e_2|^2} + e_2$, $e_2 = (0, 1)$.

We need the following notation:

$$D_1^\gamma = \{x \in D^+ | \frac{\pi}{2} - \gamma \geq \theta \geq 0\},$$

$$D_2^\gamma = \{x \in D^+ | \frac{\pi}{2} \geq \theta \geq \gamma\},$$

where θ is the usual angular variable. Next, denote

$$Q(x_1, x_2) = \{y \in D^+ | y_1 \geq x_1, y_2 \geq x_2\},$$

$$\Omega(x_1, x_2, t) = \frac{4}{\pi} \int\limits_{Q(x_1,x_2)} \frac{y_1 y_2}{|y|^4}\omega(y, t)dy.$$

Before we prove Theorem 4.11, we need the following key lemma.

Lemma 4.12. *Assume ω_0 is odd in x_1. Fix small $\gamma > 0$. Then there exists $\delta > 0$ such that*

$$u_1(x, t) = -x_1\Omega(x, t) + x_1 B_1(x, t), \ |B_1| \leq C_\gamma \|\omega_0\|_{L^\infty}, \ \forall x \in D_1^\gamma, |x| \leq \delta, \quad (4.21)$$

$$u_2(x, t) = x_2\Omega(x, t) + x_2 B_2(x, t), \ |B_2| \leq C_\gamma \|\omega_0\|_{L^\infty}, \ \forall x \in D_2^\gamma, |x| \leq \delta. \quad (4.22)$$

Remark 4.13. *We will see that in the certain regimes the first term on the right hand sides in (4.21) and (4.22) is truly the main term. Then, in the main term, the trajectories of fluid motion near the origin are pure hyperbolas. Also, note that the singularity in Ω is capable of creating exactly $\sim \log x_1$ behavior, akin to Bahouri-Chemin example, as the support of vorticity approaches the origin.*

Proof. We will consider the case of u_1; the derivation for u_2 is similar. Due to symmetry, $u(x) = \frac{\nabla^\perp}{2\pi} \int_{D^+} \ln(\frac{|x-y||\tilde{x}-\bar{y}|}{|x-\bar{y}||\tilde{x}-y|})\omega(y, t)dy$, where $\tilde{x} = (-x_1, x_2)$. Fix $x = (x_1, x_2) \in D_1^\gamma$ and take $\Gamma = 100(1 + \cot\gamma)x_1$. Since $x \in D_1^\gamma$, we have $100|x| < \Gamma$.

First,

$$\begin{aligned}\frac{\nabla^\perp}{2\pi} \int_{B_\Gamma(0)} \ln(\frac{|x-y||\tilde{x}-\bar{y}|}{|x-\bar{y}||\tilde{x}-y|})\omega(y, t)dy &\leq C\|\omega_0\|_{L^\infty} \int_{B_{2\Gamma}(0)} \frac{dy}{|x-y|} \\ &\leq C\|\omega_0\|_{L^\infty} \int_0^{2\Gamma} \frac{1}{s} s ds \\ &\leq C\|\omega_0\|_{L^\infty} x_1.\end{aligned}$$

In the rest of integration region, we have $|y| > 100|x|$. Note that

$$\pi G_D(x,y) = \frac{1}{4}\Big(\ln(1 - \tfrac{2xy}{|y|^2} + \tfrac{|x|^2}{|y|^2}) - \ln(1 - \tfrac{2x\bar{y}}{|\bar{y}|^2} + \tfrac{|x|^2}{|\bar{y}|^2}) \\ - \ln(1 - \tfrac{2\tilde{x}y}{|y|^2} + \tfrac{|x|^2}{|y|^2}) + \ln(1 - \tfrac{2\tilde{x}\bar{y}}{|\bar{y}|^2} + \tfrac{|x|^2}{|\bar{y}|^2})\Big).$$

Observe that $\ln(1+s) \sim s - \frac{s^2}{2} + O(|s|^3)$ for small s. Moreover, one can verify that $\frac{\bar{y}_1}{|\bar{y}|^2} = \frac{y_1}{|y|^2}$, $\frac{\bar{y}_2}{|\bar{y}|^2} = 1 - \frac{y_2}{|y|^2}$. Then, after a computation, we obtain

$$\pi G_D(x,y) = -\frac{4x_1x_2y_1y_2}{|y|^4} + \frac{2x_1x_2y_1}{|y|^2} + O\left(\frac{|x|^3}{|y|^3}\right). \tag{4.23}$$

This asymptotic expansion can be differentiated, and we get

$$\pi \frac{\partial G_D}{\partial x_2}(x,y) = -\frac{4x_1y_1y_2}{|y|^4} + \frac{2x_1y_1}{|y|^2} + O\left(\frac{|x|^2}{|y|^3}\right). \tag{4.24}$$

Notice that

$$\int_{D^+\bigcap B_\Gamma^c} \frac{y_1}{|y|^2} dy \le \int_\Gamma^2 \frac{1}{s} s ds \le 2,$$

$$|x|^2 \int_{D^+\bigcap B_\Gamma^c} \frac{1}{|y|^3} dy \le |x|^2 \int_\Gamma^2 \frac{1}{s^3} s ds \le C|x|^2\Gamma^{-1} = C(\gamma)x_1,$$

$$\int_{x_1}^2 \int_0^{x_1} \frac{y_1y_2}{|y|^4} dy_1 dy_2 \le \int_0^{x_1} \frac{y_1}{y_1^2 + x_1^2} dy_1 \le C(\gamma),$$

$$\int_{x_1}^2 \int_0^{C(\gamma)x_1} \frac{y_1y_2}{|y|^4} dy_2 dy_1 \le C(\gamma).$$

Combining all our estimates together, we get (4.21). Similarly, we can prove (4.22) for $x \in D_2^\gamma$.

□

With this main lemma in hand, exponential growth of gradient of the vorticity is easy to obtain.

Set $\omega_0 = 1$ for every $x \in D^+$ except for $x_1 \le \delta$. Then for every t, $\Big|\{x \in D^+ | \omega(x,t) \ne 1\}\Big| \le 2\delta$ (since $\omega(x,t) = \omega_0(\Phi_t^{-1}(x))$, and Φ_t^{-1} is measure preserving). Then, provided that $|x| \le \delta$, $\Omega(x,t) \ge C\int_{C\sqrt{\delta}}^{C}\int_{\frac{\pi}{6}}^{\frac{\pi}{3}} \frac{\sin 2\theta}{s} ds d\theta \ge C\log\delta^{-1}$. We can choose δ so that $C\log\delta^{-1} > 100C(\gamma)$, with $C(\gamma)$ the constant in (4.21). For any characteristic on ∂D with a starting point (x_1,x_2) satisfying $x_1 \le \delta$, we have $\frac{d}{dt}\Phi_t^1(x_1,x_2) \le -c\log\delta^{-1}\Phi_t^1(x_1,x_2)$ for some $c > 0$.

Now we are going to deal with double exponential growth. The construction is qualitatively different, and has to be essentially non-linear. We have to derive an estimate on growth of $\Omega(x,t)$ in time due to advance of the unit vorticity towards origin.

It is not clear why such advance has to be at all orderly and controllable; depletion of the region of high vorticity as it approaches the origin appears a distinct possibility. A key role in the proof plays a hidden "comparison principle" in (4.21). Namely, the region $Q(x_1, x_2)$ over which we integrate in the main term of the right hand side in (4.21) tends to be larger for points closer to origin. It is this feature that allows control of the scenario and proof of double exponential growth.

We will still assume

$$\omega_0 = 1, \qquad x_1 \geq \delta,$$

$0 \leq \omega_0 \leq 1$ in D^+. So from exponential growth proof, we have

$$\Omega(x,t) \geq C \log \delta^{-1} \geq 100C(\gamma), \qquad \forall |x| \leq \delta, \forall t. \tag{4.25}$$

For convenience, we also need the following notation. Take $\epsilon < \delta$, and denote

$$O_{x',x''} = \{x \in D^+ | x' \leq x \leq x'', x_2 \leq x_1\}.$$

In addition to the "front" of unit vorticity for $x_1 \geq \delta$, set $\omega_0 = 1$ on $O_{\epsilon^{10},\epsilon}$, with $\|\nabla \omega_0\|_{L^\infty} \sim \epsilon^{-10}$. Furthermore, define

$$\bar{u}(x_1,t) = \max_{(x_1,x_2)\in D^+,\, x_2 \leq x_1} u_1(x_1,x_2,t),$$

$$\underline{u}(x_1,t) = \min_{(x_1,x_2)\in D^+,\, x_2 \leq x_1} u_1(x_1,x_2,t).$$

Introduce $a(t)$ and $b(t)$ as follows: $a(t)$ solves

$$\frac{d}{dt}a(t) = \bar{u}(a(t),t), \quad a(0) = \epsilon^{10};$$

$b(t)$ solves

$$\frac{d}{dt}b(t) = \underline{u}(b(t),t), \quad b(0) = \epsilon.$$

Proof of Theorem 4.11. We first claim that $\forall t \geq 0$, $\omega(x,t) = 1$ if $x \in O_{a(t),b(t)}$. Indeed, assume not: $\omega(z,t) \neq 1$, for some $z \in O_{a(t),b(t)}$, then $z = \Phi_t(x)$ for some $x \notin O_{\epsilon^{10},\epsilon}$. Then $\Phi_s(x) \in \partial O_{a(s),b(s)}$ at some time s for the first time. However, by definition of $a(s)$ and $b(s)$, $\Phi_s(x)$ can not enter from the sides $x_1 = a(s), b(s)$ of the region. Due to the boundary condition, it also cannot enter from ∂D part of the boundary. This leaves the diagonal part of the boundary where $x_1 = x_2$. By our choice of ω_0, for all $s \geq 0$, the region $O_{a(s),b(s)}$ lies in $D_1^\gamma \cap \{|x| < \delta\}$. Then by Lemma 4.12, we have

$$\frac{\log \delta^{-1} - C}{\log \delta^{-1} + C} \leq \frac{-u_1(x_1,x_1)}{u_2(x_1,x_1)} \leq \frac{\log \delta^{-1} + C}{\log \delta^{-1} - C}. \tag{4.26}$$

We can assume that δ is small enough so that $\log \delta^{-1} \gg C$, so equation (4.26) means that $\Phi_s(x)$ can not enter through diagonal side. Together we proved the claim.

Now we look at

$$\begin{aligned}\frac{d}{dt}a(t) &= \bar{u}(a(t), t)\\ &\le -a(t)\Omega(a(t), x_2(t), t) + Ca(t)\\ &\le -a(t)\Omega(a(t), 0, t) + 2Ca(t).\end{aligned}$$

In the above computation, $x_2(t)$ is the value of the second coordinate where the maximum of u_1 is achieved (keep in mind that u_1 is negative), satisfying $0 \le x_2(t) \le a(t)$. In the last step, we used the inequality $\Omega(a(t), x_2(t), t) \ge \Omega(a(t), 0, t) - C$, which can be verified by direct computation. Similarly,

$$\begin{aligned}\frac{d}{dt}b(t) &= \underline{u}(b(t), t)\\ &\ge -b(t)\Omega(b(t), x_2(t), t) - Cb(t)\\ &\ge -b(t)\Omega(b(t), b(t), t) - 2Cb(t)\end{aligned}$$

(since $\int_b^1 \int_0^b \frac{y_1 y_2}{|y|^4} dy_1 dy_2 \le C$).

Note that since $\omega(x, t) = 1$ on $O_{a(t),b(t)}$,

$$\Omega(a(t), 0, t) \ge \Omega(b(t), b(t), t) + \frac{4}{\pi} \int\limits_{O_{a(t),b(t)}} \frac{y_1 y_2}{|y|^4} dy.$$

The last term above can be estimated as follows

$$\frac{4}{\pi} \int\limits_{O_{a(t),b(t)}} \frac{y_1 y_2}{|y|^4} dy \ge \frac{4}{\pi} \int\limits_{\epsilon}^{\frac{\pi}{4}} \int\limits_{\frac{a(t)}{\cos\psi}}^{\frac{b(t)}{\cos\psi}} \frac{\sin 2\theta}{r} d\theta dr \ge C(\log b(t) - \log a(t)).$$

Combining these results together, we obtain

$$\begin{aligned}\frac{d}{dt}\Big(\log b(t) - \log a(t)\Big) &\ge -\Omega(b(t), b(t), t) - 2C + \Omega(a(t), 0, t) - 2C\\ &\ge C(\log b(t) - \log a(t)) - 4C.\end{aligned}$$

By Gronwall's inequality,

$$\log \frac{b(t)}{a(t)} \ge (\log \epsilon^{-9}) e^{Ct} - 4Ce^{Ct}.$$

If ϵ is chosen small enough, then $\log \frac{b(t)}{a(t)} \ge (\log \epsilon^{-8}) e^{Ct}$. Since $b(t)$ is less than 1, we get $a(t) \le \epsilon^{8e^{Ct}}$. This gives double exponential growth of $\|\nabla\omega\|_{L^\infty}$. □

4.5 Towards the 3D Euler

In this section, we come back to Hou-Luo scenario for singularity formation in 3D Euler equation, and discuss one-dimensional models designed to get insight into it. We also review some of the earlier one-dimensional models, which have a long history in mathematical fluid mechanics. Let us begin by writing down the axisymmetric 3D Euler equation in cylindrical coordinates.

Assume $u(x) = u_r(r,z,t)e_r + u_z(r,z,t)e_z + u_\theta(r,z,t)e_\theta$, $\omega(x) = \omega_r(r,z,t)e_r + \omega_z(r,z,t)e_z + \omega_\theta(r,z,t)e_\theta$, where r, z, θ are usual cylindrical coordinates. The 3D axisymmetric Euler equation can be written as follows:

$$\begin{cases} \partial_t\left(\frac{\omega_\theta}{r}\right) + u_r\partial_r\left(\frac{\omega_\theta}{r}\right) + u_z\partial_z\left(\frac{\omega_\theta}{r}\right) = \partial_z\left(\frac{(ru_\theta)^2}{r^4}\right), \\ \partial_t(ru^\theta) + u_r\partial_r(ru^\theta) + u_z\partial_z(ru^\theta) = 0, \\ (u_r, u_z) = (r^{-1}\partial_z\psi^\theta, -r^{-1}\partial_r\psi^\theta), \; L\psi^\theta = \omega_\theta. \end{cases}$$

Here $L = r^{-1}\partial_r(r^{-1}\partial_r) + r^{-2}\partial_z^2$.

Away from the axis $r = 0$, axi-symmetric 3D Euler equation is very similar to the 2D inviscid Boussinesq system, describing motion of incompressible buoyant flow.

$$\begin{cases} \partial_t\omega + (u\cdot\nabla)\omega = \partial_{x_1}\rho, \\ \partial_t\rho + (u\cdot\nabla)\rho = 0, \\ u = \nabla^\perp(-\Delta)^{-1}\omega. \end{cases}$$

We will think of this equation set either on a rectangle or an infinite (in x_1) strip with $u_2 = 0$ condition on horizontal boundaries and either $u_1 = 0$ or periodic boundary conditions in the x_1 direction.

The singularity formation scenario of Hou and Luo [15] involves, when translated to the 2D Boussinesq case, an initial vorticity odd in x_1 and density even in x_1. Due to symmetry, $x_1 = 0$ serves as a separatrix of the flow for all times, and the flow has hyperbolic points where $x_1 = 0$ axis and the boundary meet. It is at these points that very fast and numerically robust growth of vorticity is observed. We see that this geometry is very similar to the 2D Euler example we discussed in the previous lecture, but now we have a more complex system. The main issue in trying to apply the 2D Euler ideas to the Boussinesq scenario is that the vorticity is no longer expected to stay bounded. This destroys the estimate of the key lemma, and makes control of the solution harder. Another layer of difficulty arises from the forcing term in vorticity equation, which can now create vorticity of both signs, potentially depleting the singularity formation. In this lecture, we will discuss some simplified one-dimensional models that have been developed in attempt to bridge the gap with three dimensions in understanding Luo-Hou hyperbolic scenario. Analysis of 1D models in fluid mechanics has a long history, and we start with a review of some earlier results.

Let us now discuss one dimensional models of 3D Euler equation, beginning with the general, rather than axi-symmetric, setting. The general 3D Euler equation in the vorticity form is given by

$$\begin{cases} \partial_t\omega + (u\cdot\nabla)\omega = (\omega\cdot\nabla)u, \quad in \quad \mathbb{R}^3, \\ u = \nabla^\perp(-\Delta)^{-1}\omega. \end{cases}$$

The most natural 1D model corresponding to the general 3D Euler equation is

$$\begin{cases} \partial_t \omega + u\partial_x \omega = \omega \partial_x u, \\ u_x = H\omega. \end{cases}$$

Here H is the Hilbert transform. This model has been considered by De Gregorio [5, 6]. De Gregorio model directly parallels the structure of the 3D Euler equation. It is reasonable to first analyze the effect of the two non-linear terms separately.

If we drop the vortex stretching term, we obtain the following active scalar transport equation

$$\partial_t \omega + u\partial_x \omega = 0, \quad u_x = H\omega.$$

As one can expect, in the absence of vortex stretching, the equation becomes globally regular. Global regularity can be proved in this case similarly to the 2D Euler argument; it is a good exercise.

On the other hand, let us omit the transport term in De Gregorio model. We arrive at the equation

$$\partial_t \omega = \omega \partial_x u = \omega H\omega.$$

This equation has been considered by Constantin, Lax and Majda [4]. Amazingly, the model turns out to be exactly solvable. Let us recall some properties of the Hilbert transform:

$$Hf(x) = \frac{1}{\pi} P.V. \int_{\mathbb{R}} \frac{f(y)}{x-y}\, dy,$$

or

$$F(Hf)(k) = -i\,\mathrm{sign}(k) Ff(k),$$

where F stands for Fourier transform.

If $f \in L_2$ then $f + iHf$ is a boundary value of an analytic function in $\mathbb{C}^+$. Then

$$(f + iHf)^2 = f^2 - (Hf)^2 + 2ifHf$$

is also an analytic function in $\mathbb{C}^+$. The real part of its boundary values is the Hilbert transform of the imaginary part of its boundary values. It follows that

$$fHf = \frac{1}{2} H(f^2 - (Hf)^2) \Rightarrow H(fHf) = \frac{1}{2}((Hf)^2 - f^2).$$

Theorem 4.14. *The solutions to Constantin-Lax-Majda model can blow up in finite time.*

Proof. Applying the Hilbert transform to the equation, we get

$$\partial_t H\omega = \frac{1}{2}((H\omega)^2 - \omega^2).$$

Let us define take $z(t) = H\omega(t) - i\omega(t)$. Differentiating in time we obtain

$$z'(t) = \frac{1}{2}z^2(t) \Rightarrow \frac{1}{z(t)} = \frac{1}{z(0)} - \frac{1}{2}t,$$

and finally $z(t) = \frac{2z(0)}{2-tz(0)}$. Hence

$$\omega(x,t) = \frac{4\omega_0(x)}{(2 - tH\omega_0(x))^2 + t^2\omega_0(x)^2}.$$

This implies finite time blow up if for some x_0 the initial data satisfies $\omega_0(x_0) = 0$, $H\omega_0(x_0) > 0$. □

Let us go back to the full De Gregorio model

$$\begin{cases} \partial_t\omega + u\partial_x\omega = \omega\partial_x u, \\ u_x = H\omega. \end{cases}$$

Is there a finite time blow-up? This question is currently open. It might be natural to guess finite time blow up; but surprisingly, the transport and vortex stretching terms appear to counteract each other.

Let us now discuss one-dimensional models developed recently specifically for Hou-Luo scenario. We start with the derivation of Hou-Luo model proposed already in [15]. Consider the 2D Boussinesq equation in the half-plane $x_2 \geq 0$, and make an additional assumption that the vorticity is concentrated in a boundary layer where it does not depend on the vertical direction x_2 :

$$\begin{cases} \partial_t\omega + (u\cdot\nabla)\omega = \partial_{x_1}\rho \quad in \quad \mathbb{R}^2_+ \times [0,\infty), \\ \partial_t\rho + (u\cdot\nabla)\rho = 0, \\ u = \nabla^\perp(-\Delta_D)^{-1}\omega, \\ \omega(x_1,x_2,t) = \omega(x_1,0,t)\chi_{[0,a]}(x_2). \end{cases}$$

As is well known, the Laplacian Green's function of the Dirichlet problem in $\mathbb{R}^2_+$ is

$$G_D(x,y) = \frac{1}{2\pi}(\log|x-y| - \log|x-\tilde{y}|),$$

where $\tilde{y} = (y_1, -y_2)$. From the Biot-Savart law we get:

$$u_1(x,t) = \int_{\mathbb{R}}\int_0^a \frac{\partial G_D}{\partial x_2}(x_1,0,y_1,y_2)\omega(y,t)\,dy_2\,dy_1;$$

$$\frac{\partial G_D}{\partial x_2}(x_1,0,y_1,y_2) = \frac{1}{2\pi}\left(\frac{-y_2}{(x_1-y_1)^2+y_2^2} - \frac{y_2}{(x_1-y_1)^2+y_2^2}\right);$$

$$\frac{1}{\pi}\int_0^a \frac{y_2}{(x_1-y_1)^2+y_2^2}\,dy_2 = \frac{1}{2\pi}\int_0^{a^2}\frac{dz}{(x_1-y_1)^2+z} = \frac{1}{2\pi}\log\left(\frac{(x_1-y_1)^2}{(x_1-y_1)^2+a^2}\right).$$

We can simplify our calculations by taking out the denominator, because there is no singularity in it. So, in the main term, we can take

$$u_1(x,t) = -\frac{1}{\pi}\int_{\mathbb{R}} \log|x_1 - y_1|\omega(y,t)\,dy_1.$$

A short computation shows that this is precisely equivalent to $\partial_x u_1(x,t) = H\omega$.

Based on the argument above, the following model of the hyperbolic point blow up scenario has been proposed by Hou and Luo [15]:

$$\begin{cases} \partial_t\omega + u\partial_x\omega = \partial_x\rho, \quad in \quad \mathbb{R}\times(0,\infty), \\ \partial_t\rho + u\partial_x\rho = 0, \\ u_x = H\omega. \end{cases}$$

The initial data ω_0, ρ_0 are assumed periodic.

Theorem 4.15. *The periodic Hou-Luo model is locally well-posed for* $(\omega_0,\rho_0) \in (H^m, H^{m+1})$ *with* $m > 1/2$.

If the solution loses regularity at time T, *we must have*

$$\int_0^t \|u_x\|_{L^\infty}\,dx \xrightarrow{t\to T} \infty \text{ and } \int_0^t \|\rho_x\|_{L^\infty}\,dx \xrightarrow{t\to T} \infty. \tag{4.27}$$

On the other hand, there exist smooth initial data for which the solution forms a singularity in finite time. In particular, the expressions in (4.27) *become infinite in finite time.*

The proof of Theorem 4.15 has been recently given in [3], and is based on an appropriate Lyapunov functional-like argument. Like in the proof of Theorem 4.11, where a hidden comparison principle played an essential role, there is a hidden positivity of certain expression that makes the proof work.

We will not discuss the proof in detail here, but we will take a look at a related, and simpler, model where the proof of blow up is more direct.

Choi, Kiselev and Yao [2] have proposed to study (4.27) with a modified Biot-Savart law

$$u(x) = -x\int_x^1 \frac{\omega(y)}{y}\,dy.$$

This law arises if one drops certain parts of the $u_x = H\omega$ law. The CKY law is also motivated by the expression for u in Lemma 4.12. The CKY rule is "almost local": if we divide $u(x)$ by x and differentiate, we get a local relationship. Thus it is easier to deal with than the truly nonlocal HL rule.

We will consider the CKY model on an interval $[0,1]$ with smooth compactly supported initial data (the periodic boundary conditions are not compatible with the CKY velocity expression).

Theorem 4.16. *Suppose* $(\omega_0, \rho_0) \in (H_0^m, H_0^{m+1})$ *for* $m \geq 2$.

Then there $\exists T < \infty$*, such that for* $0 \leq t \leq T$ *a unique solution* (ω, ρ) *in* (H_0^m, H_0^{m+1}) *exists.*

For the solution of the CKY model to lose regularity at time T, we must have

$$\int_0^t (||\nabla\rho||_\infty \quad and \quad ||\nabla u||_\infty \quad and \quad ||\omega||_\infty)\, ds \overset{t\to T}{\longrightarrow} \infty. \tag{4.28}$$

There exist initial data $(\omega_0, \rho_0) \in C_0^\infty([0,1])$ *such that the corresponding solution blows up in finite time. In particular, the expressions in (4.28) become infinite in finite time.*

Let us denote by $\Omega(x,t)$ the integral

$$\Omega(x,t) = \int_x^1 \frac{\omega(y,t)}{y}\, dy.$$

Let us define trajectories

$$\frac{d\Phi_t}{dt}(x) = u(\Phi_t(x), t), \quad \Phi_0(x) = x.$$

Lemma 4.17. *The following equality holds:*

$$\frac{d}{dt}\Omega(\Phi_t(x), t) = \int_{\Phi_t(x)}^1 \frac{\omega^2(x,t)}{y}\, dy + \int_{\Phi_t(x)}^1 \frac{\partial_x\rho(y,t)}{y}\, dy.$$

Proof.

$$\frac{d}{dt}\Omega(\Phi_t(x), t) = \partial_t\Omega(\Phi_t(x), t) + \partial_x\Omega(\Phi_t(x), t)\cdot u(\Phi_t(x), t). \tag{4.29}$$

Now

$$\partial_x\Omega(\Phi_t(x), t) = -\frac{\omega(\Phi_t(x), t)}{\Phi_t(x)},$$

and so the second term in (4.29) is equal to $\omega(\Phi_t(x), t)\cdot\Omega(\Phi_t(x), t)$. Next,

$$\partial_t\Omega(x,t) = \int_x^1 \frac{\partial_t\omega}{y}\, dy = -\int_x^1 \frac{u\partial_x\omega - \partial_x\rho}{y}\, dy$$

$$= \frac{u(x,t)\omega(x,t)}{x} + \int_x^1 \omega\cdot\frac{\partial}{\partial y}(-\Omega(y,t))\, dy + \int_x^1 \frac{\partial_x\rho}{y}\, dy.$$

The second integral in the last line equals $\int_x^1 \frac{\omega^2(y)}{y}\, dy$.

Adding together $\partial_t\Omega(x,t) + u(x,t)\partial_x\Omega(x(t), t)$ we get the result. □

Proof. Let ρ_0 be smooth, nonnegative, supported in $[1/4, 3/4]$, with $\max\rho_0 = \rho_0(1/2) = 2$, and $\rho_0(1/3) = 1$. Moreover, assume ρ_0 is increasing in $[1/4, 1/2]$, and decreasing in $[1/2, 3/4]$. Let ω_0 be smooth, nonnegative, supported in $[1/4, 1/2]$, with $\omega_0 \equiv M$ in $[0.3, 0.45]$.
Let us take x_n defined by $\rho_0(x_n) = \frac{1}{2} + 2^{-n}$, and set $x_\infty = \lim_{n\to\infty} x_n$. Furthermore, set $\Phi_n(t) := \Phi_t(x_n)$, and notice that

$$\frac{d}{dt}\Phi_n(t) = u(\Phi_n(t), t) = -\Phi_n(t)\Omega(\Phi_n(t), t).$$

We denote

$$\Psi_n(t) = -ln\ \Phi_n(t).$$

Then $\frac{d}{dt}\Psi_n(t) = \Omega_n(t)$, and by Lemma 4.17 we have

$$\frac{d}{dt}\Omega_n(t) \geq \int_{\Phi_n(t)}^{1} \frac{\partial_x\rho(y,t)}{y}\,dy \geq \int_{\Phi_n(t)}^{\Phi_{n-1}(t)} \frac{\partial_x\rho}{y}\,dy - 4 \geq \frac{1}{\Phi_{n-1}(t)}2^{-n}.$$

Here in the second step we had to estimate the contribution from the region where the derivative of ρ is negative. This can be done without difficulty as this region lies away from the kernel singularity. We leave details to interested reader.

Therefore,

$$\frac{d^2}{dt^2}\Psi_n(t) \geq 2^{-n}e^{\Psi_{n-1}(t)}.$$

Then by taking $t_n = 2 - 2^{-n}$ and running an inductive argument we can get a recursive estimate $\Psi_n(t_n) := a_n \geq \exp(a_{n-1} - 3n)$.

Inductively we can show that $a_n \to \infty$. For example, if $a_1 > 20$ then one can verify that $a_n \geq \exp\exp\exp(n-1)$. □

Acknowledgment: AK and CL acknowledge partial support of the NSF grant DMS-1412023. O.L acknowledges grants from Marie Curie FP7-IEF program, Acronym : TRANSIC. The authors would also like to thank the organizers of the Levico summer school on Fluids, Transport and Mixing for the invitation and Centro Internazionale per la Ricerca Matematica for making the school possible.

Bibliography

[1] H. Bahouri, J.-Y. Chemin, *Équations de transport relatives à des champs de vecteurs non-lipschitziens et mécanique des fluides. (French) [Transport equations for non-Lipschitz vector Fields and fluid mechanics]*, Arch. Rational Mech. Anal., 127 (1994), no. 2, 159–181

[2] K. Choi, A. Kiselev and Y. Yao, *Finite time blow up for a 1D model of 2D Boussinesq system*, Comm. Math. Phys. 334 (2015), 1667-1679

[3] K. Choi, T. Hou, A. Kiselev, G. Luo, V. Sverak and Y. Yao, *On the finite-time blowup of a 1D model for the 3D axisymmetric Euler equations*, preprint arXiv:1407.4776, to appear at Commun. Pure Appl. Math.

[4] P. Constantin, P. D. Lax, and A. Majda, *A simple one-dimensional model for the three-dimensional vorticity equation,* Comm. Pure Appl. Math. 38 (1985), 715–724
[5] S. De Gregorio, *On a one-dimensional model for the three-dimensional vorticity equation,* J. Stat. Phys. 59 (1990), 1251–1263
[6] S. De Gregorio, *A partial differential equation arising in a 1D model for the 3D vorticity equation,* Math. Methods Appl. Sci. 19 (1996), 1233–1255
[7] S. Denisov, *Infinite superlinear growth of the gradient for the two-dimensional Euler equation,* Discrete Contin. Dyn. Syst. A, 23 (2009), 755–764
[8] S. Denisov, *Double-exponential growth of the vorticity gradient for the two-dimensional Euler equation,* to appear in Proceedings of the AMS, 143 (2015), 1199–1210
[9] L.C. Evans, Partial Differential Equations, Second Edition, AMS, 2010
[10] L. Euler, *Principes généraux du mouvement des fluides,* Mémoires de L'Académie Royale des Sciences et des Belles-Lettres de Berlin 11 (4 September 1755, printed 1757), 217–273
[11] D. Gilbarg and N. Trudinger, *Elliptic Partial Differential Equations of Second Order,* Springer-Verlag, Berlin Heidelberg 2001
[12] E. Hölder, *Über die unbeschränkte Fortsetzbarkeit einer stetigen ebenen Bewegung in einer unbegrenzten inkompressiblen Flüssigkeit,* Math. Z. 37 (1933), 727–738
[13] V. I. Judovic, *The loss of smoothness of the solutions of Euler equations with time (Russian),* Dinamika Splosn. Sredy, Vyp. 16, Nestacionarnye Problemy Gidrodinamiki, (1974), 71–78
[14] A. Kiselev and V. Sverak, *Small scale creation for solutions of the incompressible two dimensional Euler equation,* Annals of Math. 180 (2014), 1205-1220
[15] G. Luo and T. Hou, *Toward the Finite-Time Blowup of the 3D Incompressible Euler Equations: a Numerical Investigation,* SIAM Multiscale Modeling and Simulation 12 (2014), 1722–1776
[16] A. Majda and A. Bertozzi, *Vorticity and Incompressible Flow,* Cambridge University Press, 2002
[17] C. Marchioro and M. Pulvirenti, *Mathematical Theory of Incompressible Nonviscous Fluids,* Applied Mathematical Sciences Series (Springer-Verlag, New York), 96, 1994
[18] N. S. Nadirashvili, *Wandering solutions of the two-dimensional Euler equation, (Russian)* Funktsional. Anal. i Prilozhen., 25 (1991), 70–71; translation in Funct. Anal. Appl., 25 (1991), 220–221 (1992)
[19] W. Wolibner, *Un théorème sur l'existence du mouvement plan d'un fluide parfait, homogène, incompressible, pendant un temps infiniment long. (French)* Mat. Z., 37 (1933), 698–726
[20] V. I. Yudovich, *Non-stationary flows of an ideal incompressible fluid,* Zh Vych Mat, 3 (1963), 1032–1066
[21] V. I. Yudovich, *On the loss of smoothness of the solutions of the Euler equations and the inherent instability of flows of an ideal fluid,* Chaos, 10 (2000), 705–719

www.ingramcontent.com/pod-product-compliance
Lightning Source LLC
Chambersburg PA
CBHW040139200326
41458CB00025B/6322
9783110571233